Computer Aided
Electronic Engineering

25·1·92

D. WONG
Tin On Court
12
£7·87

Computer Aided Electronic Engineering

W. PATRICK O'REILLY

 Van Nostrand Reinhold (UK)

First published in 1986 by
Van Nostrand Reinhold (UK) Co. Ltd
Molly Millars Lane, Wokingham, Berkshire, England

Typeset in Plantin 10 on 11pt by
Colset Private Ltd,
Singapore

Printed in Great Britain by
J.W. Arrowsmith Ltd, Bristol

British Library Cataloguing in Publication Data

O'Reilly, W. Patrick
    Computer aided electronic engineering
    1. Electronics—Data processing
    I. Title
    621.381′028′5 TK7835

ISBN 0-442-31747-6

Contents

Preface

Few books have yet been written on computer aided engineering as a subject in its own right. This, in itself, is no reason for writing any book, and it must first be established that existing texts are either incomplete, outdated, or markedly inferior to the proposed work in some substantial way — or that a broader, or a more specialized text would fill a real need.

In order to provide specialist short courses for industry, and to establish an HND course in computer aided electronic engineering at West Glamorgan Institute of Higher Education, I found it necessary to create course notes, assignments and projects in various aspects of computer aided design and manufacture. Enquiries amongst colleagues in other higher education establishments reinforced my view that a textbook concerned with all aspects of the engineering of electronics products was needed, and that the use and integration of computer-based tools for such work should be the main theme of book.

This book is intended as a study course in the application of modern electronic engineering techniques using computer technology. The level is such that the work should meet the needs of final year students on electrical and electronic engineering courses at HNC / HND level, and should provide a broad introduction to the subject for undergraduates prior to more specialized final year studies into design engineering, production engineering or management.

Practising engineers and managers in the electronics industry should find that this book provides an easily assimilated means of updating their knowledge of CAE in areas where their present responsibilities provide limited opportunity for keeping in touch with new developments.

With such a wide-ranging potential readership in mind, it has been necessary to plan the structure of the book in such a way that required information can be accessed easily. This has necessitated extensive use of sub-headings identifying the subject areas of sections of text. Inevitably there are areas of overlap between these sections and I have taken advantage of these overlaps to provide discussion of factors important to the successful integration of the various design, manufacturing and management systems which constitute CAE in its broader sense.

Chapter 1 is concerned with 'task definition'. The processes and decisions which must be considered as a new product progresses through each stage of its life-cycle from Design through Manufacture and Customer Support are considered in some depth, since these are the tasks which must be addressed by the computer aids.

Chapters 2 to 7 discuss techniques of applying computer aids to the design, manufacture and test of electronics products. In each instance a chapter devoted to techniques is followed by a complementary chapter describing current systems for applying the techniques.

Chapter 8 is devoted to the implementation and management of CAE systems for design, manufacture, test and project management via an integrated system of computer-based tools.

Chapter 9 provides a brief look at some 'emerging technologies' which might be expected to influence significantly the path of future developments in CAE. I have resisted the temptation to make Orwellian predictions of what electronic engineering might involve in the next century. Instead I have focused attention on just a few of the results of current research which appear to offer opportunities for more immediate exploitation.

Appendix 1 is a suggested list of learning objectives from which lecturers might wish to select material for incorporation into college devised BTec units at level H (levels IV and V). [Copyright on the material from this appendix only is waived and it may be reproduced without reference to the publisher.]

Acknowledgements

In preparing this book I received many helpful suggestions from colleagues both in industry and in higher education. In particular both the publishers and I would like to thank the many companies who kindly provided photographs, diagrams and other resource materials. These companies include Blundell Production Equipment, Daisy Systems, Deltest, Dyna/Pert Precima., GE-Calma, IV Products, Lattice Logic, Mullard, Plessey, Racal Redac, Siliconix, Vero Advanced Products, Wayne Kerr and Zehntel.

I also gratefully acknowledge the contributions of my colleagues Raymond J. Lewis and John Walchester who helped check the material and the examples used in this book. Last, but by no means least, I am particularly indebted to the anonymous reviewer, Dr Richard Weston of the Department of Production Engineering, Loughborough University of Technology, who painstakingly read the draft and provided invaluable advice throughout the preparation of this book.

W Patrick O'Reilly

1 The electronic engineering business

1.1 OBJECTIVES

This chapter provides an over-view of the 'life cycle' of a typical electronic product from its initial conception, through design, manufacture and customer support; and an insight into the principal business and technological processes involved. Later chapters focus on these processes in more detail, with particular emphasis on the application of computers within electronic engineering. In reading this chapter you should achieve an appreciation of the main processes involved in:

— the development of a new product
— the manufacture of semiconductor devices
— the manufacture of hybrid microcircuits
— the manufacture of printed circuit assemblies
— the testing of electronic equipment
— supporting a product 'in the field'
— controlling an electronics business.

1.2 MARKETING CONSIDERATIONS

Marketing can be described as aiming to maximize the benefits of a product to its *customers* whilst ensuring its profitability for the company [1]. Certainly, then, it is necessary to consider everything about the product from the birth of an idea to its eventual replacement due to obsolescence: a 'womb to tomb' involvement. Selling, which is just one important aspect of marketing, aims to maximize the benefits to the *company*. For most products there is a need to *define, design, manufacture, sell* and *support* by after-sales service. Normally the major source of revenue or 'positive cash flow' to the business acrues as a result of *selling*. The product definition and design stages of a project require expenditure of cash and are periods of 'negative cash flow'. Repairs under warranty may be offset by profitable supply of spares later, although towards the end of the life of a product components may become obsolescent and the company may choose to carry excess costs of spares to avoid losing customer goodwill. Thus the tail end of a product life-cycle can involve some negative cash flow. Fig. 1.1 illustrates the cumulative cash flow during a typical product life-cycle.

1.2.1   Product life-cycle trends

It is instructive to investigate life-cycle timescales and trends, particularly in the electronics industry, since to a very large extent they are closely correlated to improvements in engineering technology. This tendency has been in evidence for many years and is not restricted to the electronics industry alone[2]. Some examples of timescales are listed in Table 1.1.

The very strong message from all this is that there is a trend — dictated by market pressures and increasing intensity of competition — towards a continual reduction of timescales for all phases of a product life cycle. Organizations which succeed in such an environment do not do so by cutting their profit margins to a minimum, since this would provide minimal funds for reinvestment in new equipment and products. In the electronics industry the company which invests wisely in improved methods and equipment to handle each stage of the life cycle can benefit from high margins over costs, thus providing the finance for further improvements in performance.

1.2.2   Innovation in engineering

Innovation is the key to capturing markets. However, creativity should not be interpreted as solely the prerogative of the research scientist and the design engineer. Certainly it is easier to be competitive if one is first onto the market with a novel product. (However, *all* products have their competitors. Even the very first motor cars had

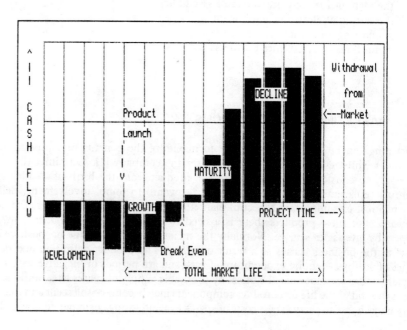

Fig. 1.1   Product life cycle
(cumulative cash flow)

Table 1.1

Invention or discovery		Date of discovery	No of years to commercial exploitation
Television		1926	10
Electronic computer		1942	4
Transistor		1947	3
Planar SIC		1958	2
Microprocessor		1970	1

Product Code	Type	Approximate launch date	Approximate market life (Yrs)
EF50	Vacuum valve	1940	30
OC72	Ge. transistor	1955	15
µA709	SIC amplifier	1961	8
4004	Microprocessor	1971	4

to compete with more traditional modes of transport. Early television receivers faced and soon overcame competition from the cinema and variety theatre.) It is also possible to achieve market domination — as has the Japanese electronics industry in the Hi-Fi, video and memory chip fields — by later entry into a market. In such circumstances, however, it is necessary to break down the barriers of supplier loyalty which have been built up by existing market leaders. Improved technology of manufacture is a major ingredient of such successes, providing lower manufacturing costs, improved quality and reliability. By such means the electronics industry has provided unparalleled improvements in value for money, as the comparisons of Fig. 1.2 indicate.

The most successful electronics companies are alert to the opportunities available as a result of combining creative design and innovative manufacturing technologies. Computer aided engineering (CAE) will promote the realization of this philosophy. CAE is not a variant or a special case of conventional engineering, but rather the logical evolu-

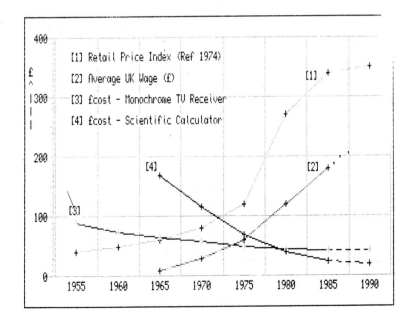

Fig. 1.2 Prices and incomes trends

tion of traditional engineering using the computer not so much as a 'tool' but rather as a material from which a complementary range of modern tools can be forged. These are tools which offer great flexibility since their characteristics can very quickly be adapted such that their performance is optimized to the needs of each particular task to which they are applied.

This is an appropriate place to investigate in some detail the tasks which CAE tools are required to undertake, prior to considering how computer technology is being applied to these tasks.

1.3 THE DEVELOPMENT PROCESS

The rapid advances of the electronics industry are underpinned by massive investments both in fundamental research into materials and manufacturing processes, and in applied research aimed at practical exploitation of technology in terms of new and better components, and new circuits and software to provide product opportunities. Many electronics companies plough back between 20% and 30% of all sales revenue into research and development.

Fig. 1.3 shows the key steps in a product development programme which it may be helpful to summarize:

Step 1. Feasibility study. Product development begins in many instances with an applied research programme or a feasibility study. The objective is to verify the practicability of novel aspects of a project either by theoretical modelling or by 'breadboarding' (building a crude hardware model) of critical areas.

Step 2. Project definition. Here a detailed plan of all tasks or 'work packages' necessary for successful completion of the project is clearly defined and an estimate is made of the likely timescale, cost and other resource requirements for each. If the product is to contain innovative features then there will be considerable uncertainty associated with a number of the work packages. An attempt is made to quantify these risks in financial terms. The basis for such an analysis is a financial appraisal of the project through the whole of its anticipated life cycle. A 'network' is drawn up showing the sequence in which tasks need to be undertaken in order to complete the project in the required timescale and with acceptable variations in the numbers of people required from month to month. On all but the simplest of projects the task of juggling work package start dates in order to obtain a reasonably level resource demand is really only practicable with computer assistance[3]. This subject is addressed in Chapter 8.

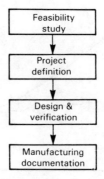

Fig. 1.3 The   development process

Step 3. Design. In this stage circuits and software are designed to provide the required performance, and the physical way in which the components of the product will fit together is decided. Circuit design usually comprises two separate activities. Firstly a circuit is postulated — often by a design engineer producing a rough sketch with descriptions of the performance of each component of the circuit — and then a mathematical analysis of the circuit forecasts the likely performance of the circuit. If tolerances in component parameters are known, then they can be taken into account to predict performance spreads in manufacturing when the product will be built in quantity. Any shortfalls in performance require modification of the circuit and re-analysis. For logic circuits such modelling techniques can provide very high confidence that the system, once manufactured, will work correctly. (One manufacturer has reported that over 97% of his gate array chips work satisfactorily first time, but this is an area where CAE is the norm rather than the exception.) Analogue circuitry, particularly if it has to work at extremely high frequencies, is more difficult to model. This is not because the calculations are necessarily more difficult, but rather because it is much more difficult to determine precise descriptions of the behaviour of the components. To reduce this risk physical prototypes, or 'A Models', may be built and the hardware fully evaluated before proceeding to the next stage.

Step 4. Manufacturing documentation. As the design nears completion, it is necessary to prepare instructions for the assembly, testing and maintenance of the product. The traditional medium for this communication with the production department is a set of drawings and specifications. In some instances it is desirable to produce a further small batch of 'B models' to verify the accuracy of this information before large investments are made in production facilities. Each of these hardware model-build programmes further delays the launch of the product and hence erodes competitiveness in the market place.

Step 5. Production start-up. Finally a manufacturing facility is established. This task in itself is a development project and may require the setting up of an industrial engineering team who must appraise available manufacturing equipment and techniques. The manufacturing engineering is equally competitive, and industrial engineers must therefore be aware of the facilities and techniques being used by their competitors. When new manufacturing techniques are involved it is not unusual for the unit cost in early production to be higher than ultimately planned, since 'learning' is involved.

1.4 CONSTRUCTION TECHNIQUES

At the present time, and indeed for the foreseeable future, most electronic equipment is produced by connecting a number of sub-assemblies and components onto a rigid or semi-rigid insulating substrate or printed circuit board (PCB). A number of such modules, installed in an enclosure, make up a functional equipment (Fig. 1.4).

1.4.1 Components

In the past equipment designers have selected from a 'menu' of items supplied by component manufacturers. Only rarely have they had recourse to creating 'custom-designed' components. While suitable ranges of discrete capacitors, resistors and in many instances inductive components have been readily available the same has not generally been true of integrated circuits. Consequently designers have often been

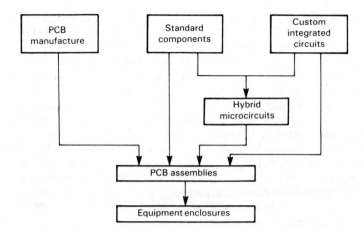

Fig. 1.4 *Process of equip-*
ment manufacture

obliged to trim their product specifications to suit the performance of available active devices. This pattern has changed radically since the late 1960s when the first custom integrated circuits were designed into original products. The benefits in terms of improved performance, increased reliability, smaller size and reduced manufacturing cost of these products were immediately appreciated, but there was a major disadvantage: the cost of designing and 'tooling up' for manufacture of custom chips made their application prohibitive for all but products with the highest production volumes.

New techniques of 'semi-custom' integration have evolved since the mid 1970s so that today equipments of quite modest production volumes can economically exploit what has come to be termed 'silicon design'. In order to fully exploit this opportunity it is essential that both the designers and the manufacturing engineers should have an understanding of the key processes, capabilities and constraints of such a realization medium. The discussion which follows relates to silicon integrated circuits in particular, although most of the stages are common to device fabrication in other semiconductor media such as gallium arsenide.

1.4.2 Semiconductor device manufacture

The scientist or engineer concerned with development of new semiconductor processes requires a detailed understanding of solid state physics and of the various processes involved in the manufacture of semiconductor materials and devices[4,5]. However, as the scale of integration increases — primarily as a result of reduction in dimensions of individual circuit features — so the performance of the components becomes much more closely linked to their physical construction and hence to the manufacturing processes. These links must be reflected in the software of the CAE systems used in the design of integrated circuits. In the following summary of the main steps in creation of an integrated circuit, manufacturing process are discussed in detail only where users of CAE systems are likely to encounter design features which need to be related to the technology for subsequent fabrication:

Step 1. Wafer preparation.   While there is no prospect of silicon shortage (about 25% of the matter of this planet is silicon) this material, as silicon dioxide or quartz in rocks and sand, is quite unsuitable for use in its natural state. It must be 'reduced' to remove the oxygen content, and purified or 'refined'. Only in the form of large single crystals can one exploit the semiconducting properties of this element. For large-scale applications a 'zone refining' process[6] is employed to reduce the concentration level of impurities to the very low value required in starting material for silicon chips. Since the required special electrical characteristics are obtained by introducing, quite intentionally, 'doping' concentrations as low as a few atoms of impurity per billion atoms of silicon, it follows that the starting material must have a purity of somewhat better than this level.

From the refined polycrystalline silicon a single crystal is grown, generally by the Czochralski vertical crystal pulling process[7]. 'CZ' pullers are currently capable of producing silicon crystals more than two metres long and weighing over 50 kg. The cylindrical crystals are sawn into thin slices or wafers which form the substrate onto or into which the active circuits will be placed. At present the semiconductor industry uses mainly 10 and 15 cm (4 and 6 inch) diameter wafers, but crystal producers have already solved the problems of 20 cm (8 inch) wafer manufacture which would give nearly double the surface area per slice. This is likely to become the next industry 'standard' during the late 1980s.

Rotation of the crystal while it is growing ensures a circular cross-section of the finished ingot. After being ground to a closely defined diameter a flat is machined onto the ingot to mark the orientation for machine handling. Diamond edged ID (inside diameter) saws are used to slice the silicon ingot into wafers of less than 1 mm thickness. These are then polished to a mirror finish using progressively finer polishing powders.

Step 2. Epitaxy. Onto the bare wafer is grown a lightly doped layer of silicon whose atoms are aligned on the same crystallographic axis as the starting material. This is termed an *epitaxial layer* and is usually produced by chemical vapour deposition (CVD) of silicon by the high-temperature decomposition of $SiCl_4$. NMOS integrated circuits, for example are built upon wafers having a p-type epitaxial layer.

Step 3. Fabrication. Fabrication normally involves the use of photographic masks which are used in the engraving of patterns into an oxide layer on the wafer surface to define various regions of the active devices. This photolithographic process is central to most chip technologies, and so it will be briefly outlined here:

(a) A layer of photoresist is spun onto the wafer surface so as to achieve very uniform coverage.

(b) A photographic mask consisting of areas of chromium on a glass substrate is brought very close to, or in some cases actually into contact with, the surface of the dry resist. The patterns of the masks are the shapes required to define the transistors and their interconnections, and it is the design of these masks which differentiates functionally between finished chips.

(c) The unmasked areas are exposed to ultraviolet light which hardens the resist. The unexposed resist is then washed off using an organic solvent.

(d) The revealed oxide regions are now removed by an acid etch to expose the underlying silicon. A second etchant removes the remaining resist so that there are now windows in the surface oxide through which either p- or n-type material can be diffused or implanted, or metal to semiconductor contacts can be made.

(e) The source and drain electrodes of the NMOS transistors are created either by diffusion or by ion implantation [8] of a group V element such as arsenic to produce n^+ doping (heavily doped n-type silicon which has low resistivity). An ion implanter bombards the exposed silicon surface with ions of the dopant. The depth of the implantation is altered by control of the ion energy, and the impurity concentration is determined by the beam intensity and the duration of the implant process. Ion implantation inevitably damages the crystal lattice structure and a period of annealing in a high-temperature oven is subsequently required. Laser annealing is sometimes used as an alternative and has the advantage of controlled depth of heating [9].

A further photoresist and etching stage removes the heavy layer of oxide at the centre of the transistor. This is replaced by a very thin oxide layer (typically only 0.1 μm thick) which is extended to just overlap the n$^+$ regions of source and drain.

Step 4. Metallization. Aluminium is used as the metallization material to inter-connect the various active devices which make up the MOS circuit. This metal adheres well to SiO$_2$ and makes good ohmic contacts with heavily doped (p$^+$ or n$^+$) silicon. The aluminium is evaporated onto the surface of the wafer and a further masking and etching process defines the areas where the metal is to remain. A small percentage of copper is added to the aluminium to reduce the effect of electromigration. This is a phenomenon whereby, as current flows through the metal, atoms move with the current until eventually an open circuit occurs at a narrow region or at a point of contact with the semiconductor. With metal film thickness of just one micron or so, narrow conductors can be subjected to very high current density in some parts of a silicon chip. This aspect of chip design is becoming all the more important as the physical dimensions of very large-scale integration (vlsi) devices move towards sub-micron geometries.

Many integrated circuits now require more than one layer of interconnections in order to avoid wasting large amounts of chip area. Heavily doped polycrystalline silicon is used for certain of the interconnections in some MOS circuits. The insulation between layers of metallization may be obtained by chemical vapour deposition (CVD) or by evaporation of a dielectric film on top of the first layer on interconnections. The design of metallization patterns is considered in more detail in chapter 3 where this subject is central to the application of computer aided design (CAD) techniques to gate-arrays. (These are pre-processed semi-custom chips whose circuit function is defined by adding a final 'commitment' or 'personalization' metal layer in accordance with the customer's specification.)

Step 5. Testing and packaging. A wafer may typically contain one hundred separate circuits each containing tens or even hundreds of thousands of transistors. A few of the circuits on each wafer are simpler patterns which can be probe tested to check the per-formance of individual devices or gates. If these simple test patterns perform properly it is likely that a high proportion of the complex chips will also operate correctly. While still in wafer form, probe testing of the chips is usually possible so that any faulty devices can be marked with an ink spot and no further processing will be carried out on them. The wafers are then scribed and separated into individual chips which are mounted into metal, ceramic or plastic packages or 'headers' prior to bonding of their contact pads to the outside world via the header leadframe. Fig 1.5 illustrates various leaded and surface-mounting packages. Performance testing and reliability 'screening' tests may then be performed on the finished devices prior to shipment.

At the next levels of assembly the individual integrated circuits are interconnected to create electronic modules and complete products. Modules or sub-assemblies generally take one of two forms: hybrid microcircuits or printed circuit assemblies.

1.4.3 Hybrid microcircuits

The small size of transistors and their ability to operate from low ('safe') operating voltages has stimulated the almost universal adoption of PCBs for equipment assembly. Whatever the scale of silicon integration there has always been a need for intercon-necting a number of devices together with other components (e.g. capacitors, inductors and connectors) whose nature is not amenable to silicon design. In the mid 1950s small,

Fig. 1.5 Surface-mounted and conventional printed circuit boards (Courtesy Mullard Ltd)

pre-assembled and tested PCBs containing miniature components in useful functional units such as counters and amplifiers, found a ready market with original equipment manufacturers (OEMs). Like other forms of modules they helped designers reduce product development 'lead times', and it was not long before these small modules appeared in encapsulated form offering improved ruggedness and environmental protection. Component manufacturers adopted and developed this philosophy using 'bare chip' transistors and small-scale integrated circuits on metal headers. Welded wire interconnections were used to bond the active devices to the tracks on the substrates of these multi-chip hybrid circuits. The benefits of very high reliability, microminiaturization and relatively low development cost (when compared with design in silicon) remain attractive features of hybrid circuit technology today.

The wire interconnections have largely, though not entirely, been replaced by metal films for connecting various parts of the circuit together. Many of the circuit components, such as resistors and small-value capacitors can also be produced in film form. Thus a modern film hybrid would normally contain conductive, resistive and insulating films as well as possibly some active devices and large value capacitors in a style suitable for surface-mounting on an insulating surface. Hybrids are very much a 'custom design' technology in which the OEM usually works in partnership with the hybrid manufacturer during the design phase. Hybrid microcircuit companies can only compete with other manufacturing technologies by making the most effective use of computer aided engineering technology and, like the semiconductor industry, they have developed some sophisticated hardware and software tools both for design and for manufacture[10]. These aspects will be considered in more detail in Chapter 2. The main features of the two predominant hybrid microcircuit technologies, thin film and thick film hybrids, are briefly outlined below:

(a) Thin film hybrids

Thin films are generally built upon glass or glazed ceramic substrates by vacuum evaporation or sputtering [11,12]. There are two basic methods of creating the required pattern in thin film technology. A metal mask can be used during deposition to obscure all of the substrate except for the areas where film is required. Masks, containing holes of the required shapes, are held closely to the substrate surface as each layer is deposited in turn. Alternatively the whole of the surface can be covered in film, whereafter the required pattern is produced by a photoresist, masking and etching process similar to that used in the fabrication of silicon circuits.

When necessary, it is possible to attach discrete components to a thin film hybrid microcircuit prior to testing and encapsulation. Some components are connected via fine wires using welding, brazing, soldering or thermo-compression bonding. Surface mounting components, such as bare ceramic chip capacitors, can be connected using metal-loaded epoxy resins to form a conductive bond[13].

(b) Thick film hybrids

Most thick film hybrids are built on alumina substrates, although other materials are occasionally employed when special characteristics are required[14]. For example beryllia — not a popular medium because of the toxic nature of beryllia dust particles — is used where improved thermal conductivity is essential.

Thick films are produced by a screen printing process using special 'inks' which, after firing in a furnace, provide the required conductive or resistive patterns. The thickness of the film cannot be controlled as closely as a deposited film, although the ratios of resistor values are fairly well controlled. Where high-precision resistors are required they need to be mechanically trimmed to the right values. The most common method of component trimming, in both thick-film and thin-film technologies, is the use of a laser. The firing of the laser is under computer control, as is the servo system which adjusts a mirror to reflect the laser beam towards the required area on the chip. A lateral laser cut provides a coarse adjustment of resistor value, while for final adjustment a longitudinal cut provides improved sensitivity[15].

Fig. 1.6 shows some typical hybrid microcircuit constructions. Active filters and radar logarithmic amplifiers are examples of circuits whose performance is critically dependent upon the values of components. Hybrid microcircuits using miniature encapsulated or bare chip active devices with passive components in film form can be trimmed for optimum performance whilst actually functional. The circuit behaviour is monitored by automatic test equipment and the results are processed by computer. In this way the laser is made to trim the circuit until its performance meets specification.

Fig. 1.6 Typical hybrid microcircuits (Courtesy The Plessey
Company Ltd).

1.4.4 PCB assemblies

At its simplest a printed circuit assembly consists of an insulating substrate carrying conducting tracks onto which are mounted and connected a number of electronic components. The PCB itself can serve some or all of several functions:

— mechanical support for the components
— low-resistance connection between circuit elements
— insulation between points to remain isolated
— electromagnetic screening
— heat spreading and dissipation

The design of a PCB assembly is therefore much more than the achievement of a successful interconnection pattern. It may be helpful to summarize the key decisions in PCB design, since some or all of these design tasks can be considered as candidates for computer assistance and are addressed in Chapter 2:

PCB materials

Some important factors in the selection of PCB insulating material are:

— cost
— mechanical rigidity
— machining properties
— dielectric constant
— power factor/loss tangent
— thermal conductivity
— thermal expansion coefficient.

Glass-reinforced epoxy board is generally used for single-sided, double-sided or multi-layer applications. However, the poor electrical characteristics of epoxy resin at micro-wave frequencies make it an unsuitable material for PCBs containing conductors the lengths of which are significant compared with the operating wavelengths. PTFE-glass boards are commonly used for applications involving striplines, directional couplers or printed filters. A more costly material than epoxy-glass, PTFE board provides very low loss and accurately controlled permittivity, but offers poorer copper adhesion (peel strength) and must be machined with great care to avoid overheating and release of toxic fumes[16].

Conductors

Copper is employed on almost all PCBs for creating the interconnection patterns. Gold plating may be necessary where track at the edge of a PCB is to be used as a wiping contact — as with mating to an edge connector. Silver or gold plating is also used in very high-frequency circuits to reduce dissipative loss due to 'skin effect'. This is a pheno-menon which results in the current penetrating only the surface skin of a conductor at high frequencies, and hence the effective resistance of the conductor can be many times that of its d.c. resistance[17]. Techniques for calculating skin effect, inductance, capacitance, characteristic impedance and inter-conductor coupling effects in PCBs are well documented[18]. For critical applications the CAD software must incorporate these 'parasitic' layout elements into the design of a circuit and verify that the perfor-mance is not excessively degraded as a result of the physical layout chosen.

The pressure for miniaturization of components would largely be wasted if traditional PCB conductor widths of several milimetres were retained for interconnections, and hence track widths and spacings of 0.2 mm and even 0.1 mm are becoming common practice in PCBs for microelectronic systems. This demands careful design of current-carrying tracks which, to avoid excessive voltage drop and overheating, must depart from this new 'norm'. For example a 0.1 mm copper track running the length of one standard dual-in-line (DIL) integrated circuit package would have a resistance of approx 50 milliohms. A current of one ampere through such a track would cause a temperature rise of around 20 °C.

The conductor pattern is produced by a photo-resist and etch process[19] from masks produced at 1 : 1 or scaled down from masters at 2 : 1 or even greater where high preci-sion is required. Traditionally master artworks, one for each layer in the PCB, are produced by attaching opaque tapes to translucent, stable plastic film. Working from a scaled layout of the board which shows the component pins and the proposed routes for their interconnection, the designer uses preformed pads and tapes of the required widths. If a mistake is made during layout, or if a modification is required to the circuit, this can involve many hours or even days work in removing and replacing tapes to achieve the 'rework'.

Artwork preparation

The process of producing artwork masters may be summarized as:

Step 1. From the circuit sketch an estimate is made of the required board size. Alternatively, if standard boards — such as Eurocards — are to be used, the circuit is partitioned into testable, functional units of the required size.

Step 2. Details are collected concerning the pin connections, package shapes, required holes and/or soldering pad sizes for each component.

Step 3. The component positions on the board are established. Some packages may contain many elements of the circuit diagram, as in the case of logic gates or operational amplifiers which are available with several functional units in a single component package. Such units on the circuit diagram must be allocated to particular packages in the board layout in a process termed 'gate allocation'.

Step 4. Routes must now be determined for the interconnections between component pins which correspond to the same node on the circuit diagram. Depending upon the complexity and density of packing on the PCB, one or more layers of conductors will be required. A point where a track from one layer must be connected through to a track on another layer is called a 'via'. Usually the routing process becomes more difficult as more connections are converted to 'routes', and the last few connections may require modification of several previously established routes before a satisfactory layout is achieved.

Step 5. Routed connections are now converted to tracks of the required widths to allow for current carrying and to keep stray inductances within required limits. Hence the circuit designer and layout designer need to cooperate over the more critical aspects of the layout.

Step 6. The finished artworks are just part of the manufacturing information for the PCB assembly. Also needed are parts lists; machining information for the bare PCB; solder resist masks which define areas at which soldered connections to components are to be made; perhaps a silk screening mask which will be used to apply component identification coding to the surface of the PCB to assist maintenance and repair; assembly instructions or diagrams; and both circuit and layout diagrams for service manuals.

Machining

Where components or connections are to be fitted through the PCB, holes must be drilled. Some grades of board material are suitable for punching, either cold or with the PCB pre-heated[20]. A router is used to machine the PCB to the required profile.

Through hole plating

Where plated-through holes are a requirement the holes are drilled before the track pattern is etched. A very fine 'seeding' layer of palladium is deposited, from solution, over the complete surface of the board, including the insides of the holes. This is followed by electroless copper plating prior to electro-plating to the required thickness of copper. A final protective coating of tin–lead plating is normally added.

Assembly

The process of assembly is a sequence of:

— select component
— identify its location on the PCB
— insert the component leads into the required holes
or
— fit the component in place if it has no hole fixings
— clinch through leads to secure component
— crop excess lead lengths
— solder (when all components are in place).

1.4.5 Testing

The important considerations of testing philosophy will be addressed in some depth in chapter 6, so here attention will be restricted to the purpose and nature of testing of electronic products during manufacture.

Testing is undertaken to ensure that an item is suitable for its intended use. The items considered here are components, sub-assemblies and completed assemblies of electronics. If, for the moment, it is assumed that the design is in all respects correct, then an item could still be unsatisfactory for one of the following reasons:

— it is the wrong item (e.g. due to incorrect coding of its packaging material);
— it has been wrongly manufactured (e.g. mounting holes drilled oversize or out of position);
— faulty components have been incorporated;
— it has been damaged since assembly (e.g. due to inadequate protection during storage or handling).

Inspection and testing are applied to reduce to an acceptable level the risk of the faulty item being passed on to the next stage of manufacture or, in the case of finished goods, to the customer.

The earlier in the assembly stage that faulty items are detected the lower is the cost of correction. Against this benefit must be weighed the costs of testing. Where finished goods are concerned, however, the true costs of delivering faulty product far exceed the warranty expenses, since reputation for quality and reliability play a vital role where follow-on orders are concerned. In practice then, finished equipments are invariably subjected to system tests in which they are exercised through several if not all of their major operating modes and the results compared against specified limits of performance. Pre-shipping tests may include a high-temperature 'soak' test or a 'burn-in' cycling of the product over its specified operational temperature range. (The subject of 'burn-in' is addressed in more detail in Chapter 6.) The object of such pre-delivery stressing of the equipment is to precipitate 'infant mortalities' of any components with incipient weaknesses.

During manufacture, testing may be applied to components (e.g. active and passive devices, bare printed circuit boards and cable assemblies) and to assemblies such as loaded PCBs (boards fitted with their components.) When testing active devices or other non-repairable sub-assemblies, one is normally only seeking a 'go/no-go' result (although the component manufacturer and reliability engineer may also wish to ascertain the *cause* of failure) whereas, if it is required to repair a faulty module or equipment one would require diagnostic information in order to determine which component(s) needed replacement.

Test results provide much more information than merely percentage 'yield' data. For example by logging and analysing test data, trends in performance of the manufacturing processes and in procurement effectiveness can be monitored. As a result, it may be found possible to relax the procurement specification of a critical and expensive component as other production tolerances which affect the product performance are brought under closer control.

1.5 CUSTOMER SUPPORT

Companies rarely achieve good customer relationships and 'brand loyalty' simply by adding on a customer service package to the end of a product plan. Customer-orientated engineering begins at the product definition stage. It includes such aspects as the aesthetics and ergonomics, or man–machine interface, of product design. (A popular term with modern electronics systems is 'user friendliness'.) Even in the very early stages of project definition it is often possible to consider, at least in general terms, a product enhancement policy which can extend the period of customer demand for the product at just the time when the cumulative cash flow of the project is most attractive — when the capital investment has been recovered and the contribution to overheads and profit is greatest.

Aspects of customer support which influence the application of CAE include:

— product information releases
— handbooks and repair manuals
— an updating system for existing users
— warranty repair / replacement
— spares provisioning
— provision of technical assistance.

A policy of 'upwards compatibility' has helped many suppliers of advanced electronic systems to retain customer loyalty. For the customer to change to a different supplier would involve substantial investment in staff retraining, whereas to replace ageing systems with a compatible next generation involves learning related only to the new features of the product.

'Industry standards' are relatively few at equipment level in the electronics business. It is not that manufacturers cannot agree that a standard is desirable, but rather that each organization has a vested interest in seeking universal adoption of its own preferred solution. When one company has, by its size or by technological lead, achieved general acceptance by the market as, for example, have IBM with their personal computer standard and Hewlett Packard with the HP–IB general purpose interface bus for automatic test equipment, then others have rapidly followed suit to provide compatible products. The HP–IB system, with but minor modifications, has been adopted as international standard IEEE-488, and customers now benefit since with generally little difficulty it is possible to build up a compatible test system with equipment items from a range of manufacturers[21].

1.6 CONTROLLING AN ELECTRONICS BUSINESS

Managing an electronics design and manufacturing operation involves product and project planning, communication, monitoring, and control of activities so as to maximize achievement against the targets within the plans. In the past it has been a prerequisite of management that knowledge and expertise acquired on previous projects — loosely termed experience — are passed on to improve performance on future ventures. To some extent learning has been achieved by trial and error, with the more successful management teams minimizing repetition of past failures. Information technology has provided the opportunity for greatly improved communications within a business, both in terms of quality of presentation and immediacy of access to up-to-date information on which to base decisions.

Human relations aspects of management skills continue to play a vital role in the motivation and hence the performance of staff in engineering organizations. This is further accentuated since, as should be clear from the following chapters, the adoption of CAE naturally leads to a more 'thinking' staff[22].

Computer aids to general management and decision making fall outside the scope of this book, although Chapter 8 will include a consideration of techniques to assist:

— bid preparation
— financial appraisal of projects
— estimating
— project planning and work scheduling
— progress motoring, control and reporting.

since such systems interact with, and will become integrated into future generations of CAE systems.

SUMMARY

Successful engineering involves much more than just sound and innovative design. The designer needs to consider in depth the operational role of the product in establishing the design criteria and specification targets. The media of realization — semiconductor chip, hybrid microcircuit or PCB — greatly influence the process of design, while the method and anticipated volume of production cannot be decided in isolation one from the other since the economics of a particular manufacturing technology depend greatly upon scale of production. Development costs for a custom integrated circuit can exceed a million pounds, but if a circuit can be integrated onto a chip successfully there is no other method of realization which can rival its low unit cost in production. The PCB is likely to remain the predominant construction technique in electronic equipments for several years to come. However, it is becoming more difficult to clearly differentiate between a PCB and a hybrid microcircuit as surface mounting miniature components and very fine track widths and clearances on PCBs result in greatly increased packing density.

The ultimate judge of whether a project has been successful is *the customer*, whose needs must be considered throughout the product life-cycle, which extends beyond the point of sale, through customer support, product enhancement and, ultimately, replacement by the next generation of product. From initial planning, through conceptual and physical design, manufacture and testing, creative engineering is required. Successful

CAE must embrace these concepts whilst exploiting the improved technical capability, productivity and profitability of computer-based tools.

REFERENCES

1. Chisnall, P.M. (1975) *Marketing — a behavioural analysis*. McGraw-Hill, London.
2. Servant-Schreiber, J.J. (Translated Steel, R.) (1979) *American Challenge*. Athenium
3. Davis, E W (Dec. 1973) Project scheduling under resource constraints. *AIIE Transactions*, **5** (4).
4. Einspruch, N.G. and Larrabee, G.B. (Eds) (1983) *VLSI Electronics*. Vols 1–8. Academic Press, New York.
5. Milnes, A.G. (1980) *Semiconductor devices and integrated electronics*. Van Nostrand Reinhold, New York
6. Walter, D.J. (1971) *Integrated circuit systems*. Butterworth, London.
7. Keenan, J.A. and Larrabee, G.B. (1983) Characterization of silicon materials for VLSI. In Ref. 4, Vol 6.
8. Mavor, J., Jack, M.A. and Denver, P.B. (1983) *Introduction to MOS LSI design*. Addison-Wesley, Reading, Mass.
9. Elliott, D.J. (1982) *Integrated Circuit Fabrication Technology*. McGraw-Hill, New York.
10. Doig, R.C. (July 1979) Computer aided design of thin film circuits. *Proc. Int. Conf. Computer Aided Design and Manf. of Electronic Components, Ccts and Systems*. IEE, London.
11. Einspruch, N.G. (series editor, 1984) *VLSI Electronics Microstructure Science*. (8 Vols). Academic Press, Orlando.
12. Walter, D.J. (1971) *Integrated Circuit Systems*. Iliffe, London.
13. Matisoff, B.S. (1978) *Handbook of Electronics Manufacturing Engineering*. Van Nostrand Reinhold, New York.
14. Towers, T.D. (1977) *Hybrid Microelectronics*. Pentach Press.
15. Crawley, M.C. (1984) Lasers and film circuits. In *Industrial Applications of Lasers*. Koebner, H. (Ed). Wiley, New York.
16. Cox, D. (April 1985) Printed circuits: mechanical cutting and drilling. *Electronic Production*, **14** (4).
17. Morton, A.H. (1966) *Advanced Electrical Engineering*. Pitman, London.
18. Gupta, K.C., Garg, R. and Chadha, R. (1982) *Computer Aided Design of Microwave Circuits*. Adtech House.
19. Scarlett, J.A. (1980) *Printed Circuit Boards for Microelectronics*. 2nd edn. Electrochemical Publications.
20. Cox, D. (May 1985) Printed circuits: routing or profile cutting. *Electronic Production*, **14** (5).
21. Poe, E.C. and Goodwin, J.C. (1981) *The S-100 and Other Micro Buses*. 2nd edn. Howard W. Sams, Indianapolis.
22. Martin, R., Downs, J.E. and Simon, M. (1985) Some human factors in effective CAD. *Computer-Aided Engineering Journal*, **2** (1), 21–3.

2 Computer aided design techniques

2.1 OBJECTIVES

In this chapter various techniques are examined for analysis, simulation and design of electronic circuits and systems and for their physical realization in the form of components, assemblies and complete equipments. The chapter is sub-divided into three parts, the first of which deals with conceptual design — the conversion of an operational requirement into a realizable and verified circuit. Part 2 addresses CAD techniques for the realization of circuits in a physical form, either as integrated circuits, PCBs or hybrid microcircuits. In the final part of this chapter, the principle of silicon compilation — conversion of an operational requirement directly into information for manufacture of a silicon chip — is considered. In reading this chapter you should gain an insight into techniques for:

- circuit analysis, design and optimization
- logic simulation
- printed circuit board and hybrid microcircuit design
- semiconductor device design
- silicon compilation.

I — Conceptual design

2.2 METHODOLOGY

In the previous chapter, the processes involved in creating a new product were outlined. Before investigating techniques in any detail we should first identify the decisions which are made in creating a design, since a CAD system must make, or at least provide options for, these decisions.

The input to the design process is a statement of requirement in the form of a technical specification, derived from consideration both of the user's operational requirement, and also of the capability of current technology. There is no unique solution to a technical requirement for an electronic circuit. If one technically feasible solution to a requirement exists then, in theory, an infinite number of other solutions can also be found[1]. Some solutions are more costly, less reliable or more critical on component tolerances than others, and we can narrow down the possible outcomes by adding constraints in these areas. For completeness it would be necessary to specify not only what the circuit should do, but also what it *must not* do. In key areas, such as the generation of or susceptibility to electrical or electromagnetic interference, it may indeed be necessary to specify a required performance level.

2.2.1 Analysis

Once a design has been realized (produced in a physical form) its actual performance is also fixed — at least for a particular set of component tolerances, operating temperatures and other environmental parameters. Thus when a circuit is investigated mathematically, using CAD techniques, a unique performance is predicted, and if this does not meet requirements then the design must be altered in some way and re-assessed. The process of calculating values of performance for a specified set of conditions is termed *analysis*, and has been the main design technique used by electronics engineers for many years.

2.2.2 Simulation

A more elaborate technique, made practicable by the availability of powerful computers, is *simulation*. Here a complete mathematical model of a circuit is constructed so that the response to a wide range of input stimuli and environmental parameters can be evaluated. Thus using a simulator is essentially the same as using the finished product . . . except of course that should the system prove unsatisfactory in any aspect it is only a matter of adjusting the parameters of the model until performance meets specification in all respects. Of course, restraints must be incorporated into the simulator to prevent the user from selecting non-realizable performance parameters (for example transmission lines having negative propagation delays). To introduce a significant change in a logic circuit simulation might typically take a few hours, compared with several months, and tens if not hundreds of thousands of pounds if the same modifications were introduced once the circuit was in production as a silicon integrated circuit.

One particularly attractive benefit of using a simulator is that the system under design can be tested against its *operational* requirement, thus reducing the risk of failure due to a misinterpretation of the end-user's needs during creation of the technical specification. In practice, of course, it must be stressed that this risk has effectively only been transferred to the stage of creation of an *accurate* model within the simulator.

2.2.3 Synthesis

The process of determining a single design solution from the starting point of a technical specification is termed *synthesis*. Mathematical techniques for synthesizing linear passive networks, such as wave filters, are well known and documented[2]. It is, of course, necessary to introduce additional constraints over and above electrical performance, and some of these 'design rules', which may relate for example to limitations of manufacturing processes, can be built into the CAD software. Solutions which violate the rules are automatically rejected. Other rules must be imposed by the user of the CAD system during the process of defining the task. For example, it may be a requirement, when producing the interconnection pattern for a particular printed circuit board, that maximum copper 'ground-plane' (the conductor connected to the framework of the equipment within which the board is to be installed) should be provided.

In practice current CAD systems for circuit design generally provide facilities somewhere between the *analysis* and *synthesis* capabilities outlined above.

2.2.4   Circuit optimization

For certain types of circuits, CAE software is available by means of which design and optimization of circuit values are achievable. The predictable nature of coupling structures (e.g. transmission lines in close proximity over distances significant compared with the wavelength of operation) has enabled engineers to produce microwave integrated circuits (MICs) which provide filtering and directional coupling characteristics on a low-loss substrate together with active devices. At such extremely high frequencies (typically 300 MHz to 300 GHz) poor prediction accuracy is obtained if components are assigned their 'lumped element' values. The propagation delay through resistors, capacitors and even their connecting conductors must be modelled using transmission-line theory in order to obtain representative results from the simulation. CAD programs for this class of work are now readily available.

The more sophisticated of these packages provide facilities for optimization of component values — including the layout of the interconnecting transmission lines — in order to maximize a chosen performance parameter. For example it may be required to achieve widest possible bandwidth in a gallium arsenide microwave amplifier consistent with input impedance mismatch remaining below a specified voltage standing wave ratio. In general, however, the process of design optimization requires continual monitoring of so many output variables that for a circuit of any complexity the only satisfactory process at present is one of repetitive interaction between the designer and his CAD system. The effects of circuit adjustments are analysed; their merits are assessed by the designer before further changes are introduced and the analysis repeated.

2.3   CIRCUIT SIMULATORS

Circuit level simulators based upon digital computers have been in common use since the 1950s. Early software of this type provided facilities for building up models of circuits from elemental components such as resistors, capacitors, inductors and voltage or current sources[3]. The task of describing anything but a relatively simple circuit was very tedious, and the creation of models which adequately described the performance of active devices was left to the user of the CAD system. In practice this meant that if new devices were selected, or if circuit conditions, frequency bands or operating environment differed from those previously simulated, then the designer had to build a suitable test circuit to evaluate samples of each type of active device in order to determine values for the components used in the equivalent circuit by which it was to be represented. The task of building up models of components remains a major source of cost and risk in circuit design today. However, where 'industry standard' devices are used or where the circuit is to be realized in integrated circuit form, very accurate device models may be obtained from the manufacturers and incorporated into a 'library' of models within the CAD system.

2.3.1 SPICE

The way in which circuit simulators work and what they can do will be illustrated by reference to one particular simulator called SPICE (*S*imulation *P*rogram with *I*ntegrated *C*ircuit *E*mphasis.) SPICE was developed at the University of California at Berkeley in the 1970s and for many years has been the most widely used circuit simulator in the electronics industry. Berkeley made SPICE freely available to industry — and indeed many universities and colleges have also benefited from this generosity. Versions have been produced for most of the popular larger scientific computers. In recent years adaptive derivatives have been produced by various software houses for running on a number of 16-bit microcomputers.

2.3.2 Facilities of SPICE

SPICE can simulate passive and active circuits, whether linear or non-linear. The size of circuit which can be analysed is only limited by the memory capacity of the computer (although analysis of very large and complex circuits can be quite time-consuming.)

Circuits may contain the following types of passive components:

— diodes, resistors, capacitors, inductors, mutual inductors and delay lines.

Active components may be described by:

— independent voltage or current sources
— voltage-controlled voltage or current sources
— current-controlled current sources

or by means of built-in models of the most widely used semiconductor devices. Commonly used models are considered in the following paragraphs:

Bipolar Junction Transistor (BJT)

The complex but accurate charge model derived by Gummel and Poon[4] is the basis for BJT models in SPICE. If the parameters of the Gummel–Poon model are not specified, SPICE defaults to the simpler Ebers–Moll model[5].

Field-Effect Transistors

Metal-oxide-semiconductor field effect transistors (MOSFETS) are modelled using an extended form of the Frohman–Grove model. A simple model, using a square-law current–voltage characteristic, is also available. Junction FETs are simulated using the Shichman–Hodges model[6].

When specifying a component in a SPICE simulation, several performance parameters may be entered. For example, the general form of a resistor specification is

$$\text{RXXXX N1 N2 VALUE [TC = TC1 [,TC2]]}$$

where:

> RXXXX is the circuit designation
> N1 and N2 are the connection nodes
> TC1 and TC2 are temperature coefficients

(*Note*: the [ ] brackets signify that these coefficients are optional. SPICE will assume their value to be zero if no data are applied.)

SPICE computes the value of a resistor at a temperature T which may be different from the nominal temperature T_n by

$$R = \text{VALUE(at } T_n) * \{1 + TC1 * (T - T_n) + TC2 * (T - T_n)\hat{}2\}$$

Thus

$$\text{R67 25 42 1K TC} = 0.002, 0.025$$

specifies a 1kilohm resistor designated R67, connected between nodes 25 and 42, and having temperature dependence determined by the two coefficients specified.

2.3.3 Types of analysis

The main categories of analysis available from SPICE are as follows:

D.C. analysis

D.C. analysis can be used to determine the change in output level of a circuit in response to change in voltage or current of a source connected between any two nodes of the circuit. SPICE calculates the d.c. operating point of the circuit by effectively replacing all inductors by short-circuit links and removing all capacitors. A d.c. analysis is automatically performed prior to a transient analysis in order to determine the initial conditions, and also before an a.c. small-signal analysis in order to establish small-signal models for the active devices. A particularly useful feature is the ability to calculate the sensitivity of a specified output value with respect to tolerances on component values.

A.C. small-signal analysis

Facilities are provided for tabular or graphical output of a.c. output values, such as voltage gain or transconductance, as functions of frequency. Electrical noise contributions from resistors and semiconductor devices are determined automatically from the operating-point equivalent circuits of the active devices. Plots of output noise level (e.g. in volts per $Hz^{1/2}$) or equivalent input noise can be produced if required.

Distortion

An assessment of distortion products may also be demanded in the small-signal analysis mode. This can be achieved with either a single sinusoidal input (for harmonic distortion) or with a two-frequency test signal (when harmonic and intermodulation distortion products are in general present.)

Transient analysis

The transient analysis facilities within SPICE allow the designer to obtain a prediction of the values of output variables as functions of time. The initial conditions of the circuit are derived automatically from the d.c. analysis. If required, a large-signal sinusoidal input may be applied, in which case it is also possible to demand a Fourier analysis of the resulting output waveforms.

SPICE offers many other facilities, including the ability to analyse circuits over a range of temperature. Unless explicitly requested otherwise, all analysis is undertaken at 300 °K (27 °C). Parameters such as device saturation currents, carrier mobility and junction potential are calculated automatically using data from the various device models.

2.3.4 Using SPICE

To illustrate the use of SPICE the simple circuit of Fig. 2.1 will be the subject of a small-signal analysis over the frequency range 100 KHz to 10 MHz. The steps in setting up the simulation are as follows:

Step 1. Having sketched the circuit, all nodes must be numbered. The common ('ground') node must be coded zero and other nodes must be allocated positive integers. It is sensible, though not absolutely necessary, to number nodes sequentially — from input to output in this case — since this eases the task of checking the input data. It is a requirement of SPICE that all nodes should have a d.c. return to ground, so if in a circuit no such path exists a very large-value resistor should be added such that it does not influence circuit performance. All nodes must, by definition, have at least two connections, with the exception of open-circuited transmission lines: these are often used in microwave circuits as matching 'stubs'[7], for example, and SPICE will therefore accept unterminated lines.

Step 2. The circuit must now be described to SPICE as a series of circuit elements and control inputs (Table 2.1) which define the model parameters and the control of the analysis.

Step 3. The output may be selected as a table of results, or as plots. In this case the results of the analysis are illustrated in the form of frequency-response graphs (Fig. 2.2).

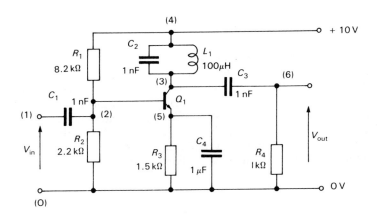

Fig. 2.1 Tuned-collector amplifier

Table 2.1 SPICE input file

TUNED COLLECTOR AMPLIFIER
INPUT LISTING   TEMPERATURE = 27.0 DEG C

```
VCC  4  0  10
VIN  1  0  AC  1
C1  1  2  1E-9F
RI  2  3  8.2K
R2  2  0  2.2K
R3  5  0  1.5K
C4  5  0  1E-6F
C2  3  4  1E-9F
L1  3  4  1E-4H
C3  3  6  1E-9F
R4  6  0  1K
Q1  3  2  5  QMOD1
.MODEL QMOD1 NPN(BF = 50 RB = 100 TF = 0.1NS CJC = 2PF)
.PLOT AC VM(6)
.AC LIN 20 100KHZ 1500KHZ
.OPTIONS ACCT LIST NODE
.END
```

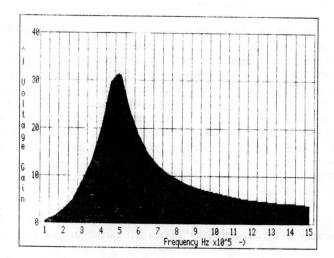

Fig. 2.2 Frequency response

2.3.5 Convergence

Key factors in the selection of a circuit simulator are the availability of device models and, where large circuits are to be designed, the speed at which the iterative process of analysis converges to a solution[8]. Indeed, there are certain cases where the algorithm used within SPICE fails to converge, particularly in transient analysis of logic circuits operating with feedback over several stages. Alternative circuit-level simulators, some using derivatives of the SPICE algorithm and others employing proprietary iterative processes, have been tailored to provide rapid, guaranteed convergence in selected types of applications and for particular device technologies[9].

2.4 LOGIC SIMULATION

A logic simulator enables a designer to construct a mathematical model of a digital circuit or system and to verify that the proposed solution, if physically constructed, would meet his needs, in terms both of operational performance (the functional and timing behaviour) and of testability. Implicit in this statement is the requirement that the simulator should provide accurate predictions of circuit performance not only when a circuit has been defined such that it can function correctly, but also under fault conditions[10].

2.4.1 Types of logic simulator

Simulators have been devised to work at various levels in logic design. Some are particularly suited to silicon chip design, while others offer improved speed in applications where systems are being built up from standard components.

Switch-level simulators

The basic modelling element (or 'primitive') used in a switch-level simulator is the active device itself, usually an MOS transistor. The device model may consist of a voltage-controlled switch varying between open-circuit and short-circuit, although simulators now exist which accurately model the characteristics of the transistors to include, for example, such parameters as 'on-resistance'.

Gate-level simulators

The familiar basic building bricks of logic circuitry — NAND and NOR gates — are the primitives of the gate-level simulator. The time taken to prepare a circuit description at gate-level is considerably reduced compared to the switch-level description. At this level functional analysis is by Boolean algebra.

Functional-level simulators

This class of design tool accepts functional block descriptions such as registers and counters. Some of these systems are capable of accepting behavioural descriptions of parts of a logic circuit which have not been designed in detail yet. Thus, for example, a designer can check out the operation, in a complete microprocessor-based system, of an arithmetic logic unit before the internal details of the memory system have been finalized.

Mixed-mode simulators

Some simulators can operate in several of the above modes. The designer can thus select the most appropriate primitive according to the nature of the work and the stage he has reached in the design. This permits a hierarchical, or 'top-down' approach to design, in which the system is described in overall performance before being broken down into progressively lower levels. Finally the circuit is built-up from low order (e.g. switch or gate-level) primitives.

2.4.2   Test-pattern generation and fault simulation

Apart from verifying that the performance of a properly functioning logic circuit would meet its required specification, a second important application of logic simulators is in the design and verification of a test specification for use when the circuit is manufactured in its final physical form. This involves selection of a set of input test vectors (a set of logic signals applied to the various inputs of the circuit) and storage of the expected responses at the outputs.

Exhaustive testing

As logic systems have increased in complexity so the task of verifying that a circuit does in fact meet specification has become progressively more difficult. Assuming that all of the logic gates within the system are there for some useful purpose, it would be necessary to create a schedule of tests, defining the application of signals to inputs and the measurement of responses at outputs, which would exercise each switching element and record a response at an output. A binary combinational logic circuit having n inputs has 2^n possible input patterns or vectors. Thus a twenty-input system has $2^{20} = 1\,048\,576$ possible test vectors. If the possibility of latches being introduced into the circuit is considered, then the situation rapidly becomes unmanageable. For a circuit with m latches and n inputs, the minimum number of patterns[11] to exhaustively test the system is 2^{m+n}

Fortunately in most cases 100% fault coverage can be obtained with just a small sub-set of these possible test patterns, and various techniques have been devised to determine, if not a minimum set, at least an efficient set of test vectors for a particular application[12].

The fault matrix

A simple example is the *fault matrix* technique, where a table is constructed containing an array of test vectors in one dimension and possible circuit faults in the other dimension. An example of such a matrix is shown in Table 2.2 for a two-input AND gate. Only faults which result in nodes becoming 'stuck at' logic levels 1 or 0 are considered.

Columns A and B define the input test pattern, while column Z contains the output results for a correctly functioning gate. The remaining six columns contain the outputs

Table 2.2

| Tests | | | Results of tests | | | | | |
A	B	Z	A @ 0	A @ 1	B @ 0	B @ 1	Z @ 0	Z @ 1
Test 1 0	0	0	0	0	0	0	0	1
Test 2 0	1	0	0	1	0	0	0	1
Test 3 1	0	0	0	0	0	1	0	1
Test 4 1	1	1	0	1	0	1	0	1

Table 2.3

Test	Results of tests A @ 0	A @ 1	B @ 0	B @ 1	Z @ 0	Z @ 1
Test 1						1
Test 2		1				1
Test 3				1		1
Test 4	1		1		1	

which would be obtained from this circuit under each of the possible single-fault conditions. (A stuck-at-0, A stuck-at-1, etc.) Thus TEST 1 involves applying 0s to both inputs, when a correctly functioning gate will give a 0 at the output z.

The table can be re-drawn to show conditions under which an abnormal output will be detected. This can be done by modulo 1 addition of each column with the 'correctly functioning' results. This is shown in Table 2.3, from which it is apparent that all possible single faults can be trapped using tests 2, 3 and 4. Test 1 is redundant and may be omitted, giving a saving of 25% in this case compared with exhaustive testing.

Path sensitization

Although this simple technique does produce a smaller test vector set than exhaustive testing, it still results in an unnecessarily large set when circuits of any complexity are involved. For logic circuits having large numbers of inputs, it is more practicable to apply test vectors and to note the output responses in the presence of stuck circuit nodes. The objective is to seek a set of vectors which together exercise all, or at least a high proportion, of the circuit nodes. A vector which allows the presence of a fault to be detected at an output point is said to have created a *sensitized path* from the faulty node to the output test point. These path sensitization techniques are thus aimed at relating measurable output responses to particular internal faults. Many faults remain undetectable if only a single path is sensitized, and so algorithms have been devised to assist in the identification of multiple paths which must be sensitized[13,14].

Logic simulators may, of course be used to investigate a design whether it is to be realized ultimately as a hybrid microcircuit, as a printed circuit board assembly or as a semiconductor device. In the latter case it will be practicable to test the finished chip only via connections to its external terminations[15]. While a test pattern *could* be devised to exhaustively exercise such a circuit, in practice this becomes unacceptably time-consuming as we move through large scale to very large scale integration, since gates deeply embedded within the circuit are inherently less testable than those near to external terminations. The testability of a gate is a function of the controllability of its inputs and the observability of its outputs — determined by how many other gates are connected in the driving or output paths respectively. Thus where no other gates are connected in either path, a single test is sufficient to identify whether or not the gate is functioning correctly. A test pattern is therefore selected which will detect a high proportion of possible single faults.

The primary purpose in testing a logic system is generally to obtain an indication of the presence or absence of faults. For non-repairable components such as integrated circuits this may be all that is required, whereas it may be possible, with further processing of the test results from a printed circuit board, to determine the nature and location of the fault. For many types of logic circuits it has been found that satisfactory results are obtained in practice if the test pattern provides a high degree of fault coverage for 'stuck at' faults when a circuit node is permanently held at logic 0 or logic 1. This assumption is justified for PMOS and NMOS technologies, although there is evidence that the ability to model 'stuck open-circuit' faults in CMOS logic provides a more realistic forecast of fault coverage[16].

2.4.3 Using a logic simulator

To illustrate the principle of simulating a logic system a description of the circuit of Fig. 2.3 is entered into and run on a typical logic simulator, in this case THEMIS, an event-driven mixed-level simulator from PRIME CAD/CAM.

Table 2.4 is the input file in which a NOR gate is defined before the circuit of Fig. 2.3 is described using three of these NOR elements. Table 2.5 contains the definition of the input test patterns (input waveforms), and Fig. 2.4 illustrates the input and output waveforms in graphical form. In this example the waveforms have been reproduced in hard-copy form using a printer.

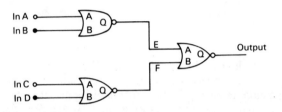

Fig. 2.3

Table 2.4 THEMIS input file

THMS>	CREATE COMPONENT NOR
COMP>	ADD PINS
PINS:	A INPUT
PINS:	B INPUT
PINS:	Q OUTPUT
PINS:	
COMP>	ADD BOOLEAN
BOOL>	EQUATION
EQTN:	Q = ~(A | B)
EQTN:	
BOOL>	DELAY VARIABLES
DLY:	TPLH = 2:4:6
DLY:	TPHL = 2:4:6
DLY:	
BOOL>	END BOOLEAN
COMP>	END COMPONENT
THMS>	CREATE COMPONENT DEMO
COMP>	ADD PINS
PINS:	INA INPUT
PINS:	INB INPUT
PINS:	INC INPUT
PINS:	IND INPUT
PINS:	OUTPUT OUTPUT
PINS:	
COMP>	ADD NETWORK
NETW>	ADD ELEMENTS
ELMT:	N1:(NOR) A = INA B = INB Q = E
ELMT:	N2:(NOR) A = INC B = IND Q = F
ELMT:	N3:(NOR) A = E B = F Q = OUTPUT

Table 2.4 *contd*

```
ELMT:
NETW>   END NETWORK
COMP>   END COMPONENT
THMS>   LINK
*NOTE:   Link started.
*NOTE:   Link of nets and propagation of float values started.
*NOTE:   Propagation of initial values started.
*NOTE:   Deallocation of unneeded structures started.

        Number of elements by nesting levels
            TOP LEVEL          =       8
        Number of primitives for all levels
            INPUT CONN-PINS    =       4
            OUTPUT CONN-PINS   =       1
            BOOLEANS           =       3

*NOTE:   Link finished.
```

Table 2.5 THEMIS input definition and control file

```
SIM>    INPUT WAVEFORM
SIM>    WAVEFORM
WFM:    INA 0 = 0 300 = 1 150 = 0 600 = 1
        300 = 0
WFM:    INB 0 = 0 *(70 70)
WFM:    INC 0 = 0 600 = 1 1000 = 0
WFM:    IND 0 = 0 130 = 1 50 = 0 500 = 1 100 = 0
WFM:
SIM>    DEF MONITOR 1
MON:    INA
MON:    INB
MON:    INC
MON:    IND
MON:    OUTPUT
MON:
SIM>    WHEN
*NOTE: Reference number 2 assigned
WHN:    INA
WHN:    INB
WHN:    INC
WHN:    OUTPUT
WHN:    [SAMPLE 1; CONTINUE]
SIM>    STEP 3000
```

2.5 Hierarchical design

The principle of top-down definition of a system followed by 'bottom-up' design may sound very straightforward. In actual practice such an approach may never converge on a satisfactory design solution, and it is frequently necessary to approach the design from both of these directions in an iterative process, seeking to meet in the middle. Possible subsystem solutions are analysed in terms of realizability, performance, testability and cost. Poor options are rejected in favour of more cost-effective solutions. Thus system partitioning cannot be established in isolation. At each level it is necessary to investigate the lower-level implications.

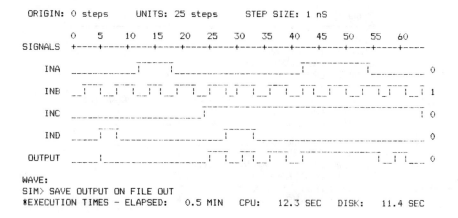

Fig. 2.4 Input and output
                    waveforms

II — Physical design

2.6 SELECTION OF DESIGN MEDIUM

Several factors need to be considered when selecting a medium (e.g. custom-designed integrated circuit, hybrid microcircuit or PCB assembly) for hardware realization of an electronic circuit as part or the whole of a product. These factors include:

— The complexity (or scale) of the circuit.
— Available technologies capable of providing the required circuit performance. This is of particular significance in the case of semiconductor devices where performance characteristics are very closely linked to the manufacturing technology.
— Timescale within which the design must be put into production.
— Weight and volume constraints of the finished product. These are important considerations for airborne and particularly for space-craft installations.
— Quantity in which the circuit is to be manufactured.
— Price acceptable in the market place. This is linked to the value a customer is likely to place upon reliability

With regard to domestic electronic systems, while much is voiced about quality and reliability, in practice few customers seem prepared to pay a high premium for a design which is likely, because of the medium of realization, to offer substantially above average reliability. This is perhaps partly related to the already high reliability of most domestic electronic products compared with mechanical or electro-mechanical equipment such as the motor car or vacuum cleaner. A second consideration is the low cost of failure of most of these products: temporary loss of television or radio reception until a repair is effected, for example.

In contrast, the risks of failure of a cardiac pace-maker, a missile early-warning system or the guidance system on a space-craft must be minimized, and a substantial cost premium would certainly be acceptable in the latter two categories if it could be shown that a significant increase in reliability could be provided by selecting the appropriate technology for manufacture of the circuitry.

Table 2.6 Typical costs and timescales

Technology	Relative chip area	Prototypes Development time (weeks)	Development cost (£K)	Quantities for cost-effectiveness (units)
PCB	—	2–3	2–4	1 +
Hybrid circuit	—	3–5	3–6	20 +
Gate array	1.5–3	4–10	5–25	250 +
Standard cell	1.2–1.8	6–20	20–80	5000 +
Full custom	1	20–70	50–200	100 000 +

Provided that the active devices are operated within their voltage, current and temperature ratings, integrated circuits provide very much higher reliability than hybrid microcircuit or PCB assemblies[17]. It follows that greatest system reliability is obtained by using the highest possible scale of integration, whether by using standard or custom devices. Table 2.6 illustrates typical design and manufacturing timescales and costs for a logic system produced in the form of an integrated circuit, in hybrid microcircuit form, and as a PCB. In each case the timescale and cost of designing the circuit itself are omitted. These figures are offered only as a general guide to realization costs since, for example, the amount and nature of production tests vary considerably according to the precise nature of the circuit. Testing costs can constitute a very high proportion of the total costs of production of certain types of circuits.

Each of these technologies — or combinations of them — are commonly employed in modern electronic products. Various surveys and predictions suggest that the use of hybrid microcircuit assembly techniques and of semi-custom chips will continue to increase and that by the early 1990s the majority of circuits will be realized in semi-custom form[18]. It seems appropriate therefore to begin a survey of CAD techniques with semi-custom integration before investigating the application of CAD to other commonly used forms of construction.

2.7 SEMI-CUSTOM INTEGRATED CIRCUITS

For many years the designers of electronic systems were entirely dependent upon semiconductor manufacturers for the types of integrated circuits from which their systems would be built. Of course, a very large volume requirement, such as television circuitry, received special attention from many semiconductor houses, but the cost of full custom design prevented the exploitation of the size, weight and reliability benefits of silicon design for all but mass-production or certain military products.

Today the facilities of semi-custom integration have brought silicon design costs within the bounds of feasibility for an increasing number of products[19]. In most cases, if total production of a few thousand is anticipated, then it is likely that a semi-custom integrated circuit will prove viable.

The uniqueness of a design in silicon is also an important commercial consideration. It would take a competitor much longer to copy the key features of a silicon chip than it would for him to produce a compatible printed circuit board.

The rapidly expanding market for application-specific integrated circuits (ASICs) stems primarily from the development of low-cost design and manufacturing techniques for these special purpose devices. While optimized custom design of a silicon integrated circuit remains for the present a very costly procedure, there are two commonly used semi-custom processes aimed at reducing these costs and hence bringing silicon design within the realms of viability for an increasing range of OEMs. These semi-custom integrated circuit technologies are *standard-cell arrays* and *gate arrays*.

2.7.1 Standard-cell arrays

Standard-cell technology is aimed at reducing the design costs associated with custom integration. The chip designer makes use of both the electrical and physical design details of functional blocks (e.g. adders, registers, etc.,) from a primitive-cell library. The cells, which are used as 'building bricks' for many ASICs, thus have well-proven performance characteristics and defined geometric shapes. To create a new chip it is necessary to place these cells upon a grid and to design a suitable pattern of interconnections between the cells. The metallization pattern within the cell itself is already defined, although it may be possible to route new conductors within the cell boundaries. A typical standard-cell chip layout is illustrated in Fig. 2.5.

An important objective during the placement and routing phase is to minimize the usage of silicon area. Since the cells themselves are fixed in area, this can be achieved only by minimizing the width of the wiring channels between cells. Other design constraints may also have to be imposed. For example it may be essential to limit the length and hence the capacitance loading on a particular line or clock signal in order to keep total delay within prescribed limits.

Once the circuit design has been completed in this way the polysilicon and metal wiring resistances and capacitances can be calculated and fed back into the simulator to provide a more accurate prediction of circuit performance[20].

It should be noted that a set of manufacturing masks for a standard-cell integrated circuit are unique to that particular device. At every stage through manufacture the wafers are dedicated to this specific application.

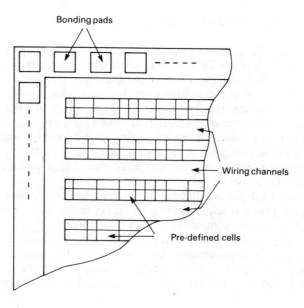

Fig. 2.5   Standard cell array

2.7.2 Gate arrays

A gate-array device achieves economies of scale in manufacture by going through most processes in a 'non-personalized' (non-committed) state. The array manufacturer prepares and holds stocks of processed wafers requiring only the addition of metal interconnections to create ASICs. The patterns on the chips are arrays of logic gates or, in some cases, more complex pre-wired functional primitives.

The task of producing the personalized metal mask patterns can be undertaken either by the device manufacturer or by the OEM. In the latter case the OEM may rent time on the chip manufacturer's CAD system or, if he intends to develop several semi-custom chips, it may be economic for the OEM to purchase the gate-array design software in a format suitable for running on his own CAD system. Many engineering workstations and PCB CAD systems are able to run the specialist software for gate-array design, the output being in a form suitable for directly driving mask-making equipments.

Gate arrays are available in various semiconductor technologies with from a few hundred to over ten thousand gage-equivalents on one chip. Emitter-coupled logic, while heavy on current consumption, offers delay times of typically 300 picoseconds per gate, and can switch at rates of several hundred MHz. When this speed is not essential CMOS has emerged as the dominant vlsi technology, providing delay times of typically 1 or 2 nanoseconds and switching rates up to 100 MHz.

It is inevitable that a circuit using the gate array approach will be less efficient in utilization of silicon area than either standard-cell or optimized full-custom form, for two main reasons. Firstly, it is necessary to select the uncommitted array containing the next available number of gates *above* the requirements of the circuit to be realized. Thus a circuit requiring say 1670 gates may have to be built from a 2000-gate array, leaving a substantial number of empty cells on the finished chip. Secondly, the wiring channels, being pre-defined, must be sufficiently wide to cope with most anticipated interconnection requirements. This means that in many gate-array ASICs the channels are wider than absolutely necessary. In practice the wiring channels are typically set so that at 80% cell utilization the autorouting program, which automatically determines routes for the interconnections, will succeed unassisted in routing most circuits[21].

The break-even point in the choice between standard-cell and gate-array approaches for a circuit depends upon many factors including complexity and choice of fabrication technology. For CMOS arrays in 1985, production quantities of less than 25 000 usually proved less costly in gate array form rathern than as standard-cell devices. The trend, however, is for a more rapid reduction of standard-cell array costs as the capability of CAD software continues to improve. Thus, while gate-array sales greatly exceeded the sales of standard-cell chips in the first half of the 1980s, a crossover around the turn of the decade appears likely[22].

The vast majority of semi-custom devices are logic circuits, although arrays containing both linear and logic elements are also available and can be used, for example, to create complete processing systems including input/output circuits[23].

2.8 PLACEMENT AND ROUTING ALGORITHMS

When first introduced, the scale of integration of gate arrays — up to a few hundred gates on a chip — was such that manual layout techniques were quite feasible. Since the early 1980s, when use of this technology by OEMs increased rapidly, a range of CAD software tools for physical design has evolved. This has been partly in response to the demand by increased numbers of users but also, and perhaps mainly, because the task of manually designing metal masks for chips of 10 000 gates or more is a daunting prospect by any other means.

2.8.1 Placement

Several configurations and semiconductor technologies are used for gate arrays, and the layout problems differ somewhat. However, they have one key feature in common: the underlying structure is fixed, so that the algorithm for placement of cells must work within these constraints. This means that it is realistic to consider the task of placement separately from that of routing the interconnections. For standard-cell integrated circuits the space left between cells can be adjusted according to the density of interconnecting conductors in that region. This results in more efficient use of silicon area provided the CAD system has the flexibility to consider placement and routing as interrelated decision processes.

Some gate-array systems use cell libraries which can be constructed from complete rows (or columns) of the array. The components can then be allocated to particular rows on the basis of connectivity, and finally individual cells can be placed within the row. The centre of the array thus resolves to a series of straightforward 'linear' placement decisions for which fast algorithms exist. The input/output (I/O) cells, which are located around the periphery of the chip, cannot be tackled in this way, but there are only small numbers of these to be connected and so they can, if necessary, be placed manually.

Not all gate arrays are so simple in structure, however. In some cases the pre-wired elements are larger functional blocks or 'macros' which may cover more than one row. For devices with non-regular structures a truly two-dimensional placement algorithm is required. The space, or 'resource' available for routing is still fixed, of course, so that placement must be made such as to avoid excessive routing congestion. The normal approach is to prepare a general 'floorplan' of the chip setting locations for highly interconnected areas of logic. These are then interchanged with one another, and with empty cells, until acceptable wire length and routing congestion are achieved. A trial routing can then be run with the option of returning to further optimize the placement if required. This type of general placement algorithm, while much slower than the row placement system previously described, has the advantage that it handles the I/O cells in exactly the same way as the internal logic cells.

2.8.2 Routing

The majority of large gate arrays use two layers of metalization for routing of connections. One layer is used predominantly for vertical runs and the other for horizontal runs of conductors. The deeper layer is used also for intra-cell connections, so that the patterns for part of this layer are dictated by the chip manufacturer. Occasionally manufacturers add a third layer of metalization for power lines, but again this layer is not accessible for customization.

Maze routers

From routing considerations, the most difficult technology is single-layer metallization where, apart from polysilicon underpasses — which can be either fixed in position or, at increased cost, may be programmable — all connections must be made via a single metal

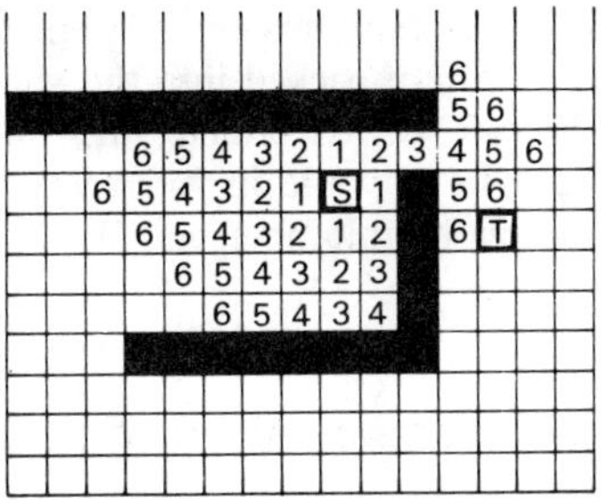

Fig. 2.6   Maze routing

mask. The solution normally adopted for routing of single-layer-metal gate arrays is based on a 'maze' algorithm[24]. Here the concept of 'costs' for a route is established, relating to the total wire length, the number of vias (connections between the metal layer and the polysilicon underpasses) and corners in the route. A search is made moving out radially from one connection along all possible routes (Fig. 2.6) while keeping a record of the cost of each path. If the other connection is reached, then the conductor track is traced back and optimized in terms of usage of vias. If a track can be found by the maze router it will always be the least-cost path. These types of maze searchers demand large amounts of computer memory and are relatively slow in execution.

Line search routers

Most common are two-layer gate arrays in which one layer allows mainly vertical connections (for example between vertical bus lines) while the other is used for crossing horizontal power buses. Line-search algorithms have been developed to provide automatic routing of these types of arrays[25]. Starting at each end of the desired connection, sets of lines are sent out in both vertical and horizontal directions. If two lines, one from each end of the connection, succeed in making contact, then a path has been found and can now be optimized to minimize vias, corners, etc. Some of these routers, whilst capable of entirely atomatic operation, provide facilities for pre-routing critical tracks.

Mode-swapping routers

There can be no complete guarantee of routing success for a gate array, particularly if the cell occupancy is high (say above 80%), and it is useful to be able to interact with the CAD system in the event of failure of the autorouter in order to move existing connections and to create space for inserting missing routes manually. The autorouter can then be re-run in a further attempt to complete the task. Some gate array design systems, for example the GARDS suite from Silvar-Licso, automatically switch to a maze algorithm when the line router stops, having rapidly completed most of the paths[26]. Thus the user gets the benefits of high speed associated with the line router, and efficient use of all available routing resource by the 'clean-up' maze router.

Channel routers

An alternative to the line router, and one well suited to standard-cell designs, is the channel router[27]. In a two-stage process, global routing first assigns portions of a desired connection to various channels and sets up entry and exit points on the channels. Then the detailed interconnections are packed into the allocated channels. If routing fails the width of channels can be expanded until all the tracks have been accommodated. This would not be possible in gate-array technology, since channel width is predetermined by the row or column spacings.

2.9 CAD FOR PCBs AND HYBRID MICROCIRCUITS

The main steps in physical design of an electronic equipment are:

— partitioning the system into physical sub systems
— assigning these subsystems to individual PCBs
— achieving layout and routing of the subsystems

2.9.1 Module partitioning

Partitioning is usually undertaken, at least initially, on the basis of functionality, so that meaningful tests can be carried out on each subsystem. However, note must be taken of the need to make effective use of package area, particularly where PCBs of a standard size are to be used. Finally, the number of connections into or out of a subsystem must also influence partitioning to some extent. When designing a PCB it may be the edge connector which limits the number of connections to the rest of the system, while for a custom-designed chip increasing the number of connections may mean changing to a much more costly package type. The process of system partitioning is particularly difficult to automate for these reasons, and manual intervention is generally necessary.

2.9.2 Interactive design of PCBs

The use of a CAD system for PCB design will be illustrated by reference to the key steps in the layout of a simple circuit (Fig. 2.7) using a Racal Redac Cadet-C PCB design system.

Step 1.   The circuit must be described to the CAD system. In this instance a textual form of description, or 'net list' has been used (Table 2.7). An alternative approach would be to make use of a schematics capture package to input the circuit diagram part of the initial data in graphical form. The initial data file for this example contains details of PCB dimensions, component references (to relate the parts to physical dimensions, pin locations and pad requirements stored in the 'component library' of the CAD system) and the connectivity of the circuit (a point-to-point wiring list).

Step 2.   Upon entering the initial data the CAD system displays the PCB outline with its edge connector, to which a series of lines representing connections radiate from a 'pile' of components at the bottom left of the screen (Fig. 2.8). These components must now be positioned on the PCB. This process is termed *placement*. A stylus moving on a graphics tablet (described in Chapter 3) is used to select and move component outlines across the screen into their proposed locations on the board. The designer signifies his satisfaction with the initial placement by depressing the CONFIRM key on the system keyboard. This initial placement is crucial, since it determines the quality of the layout, both in terms of efficient use of board area and also in minimizing interconnection tracks

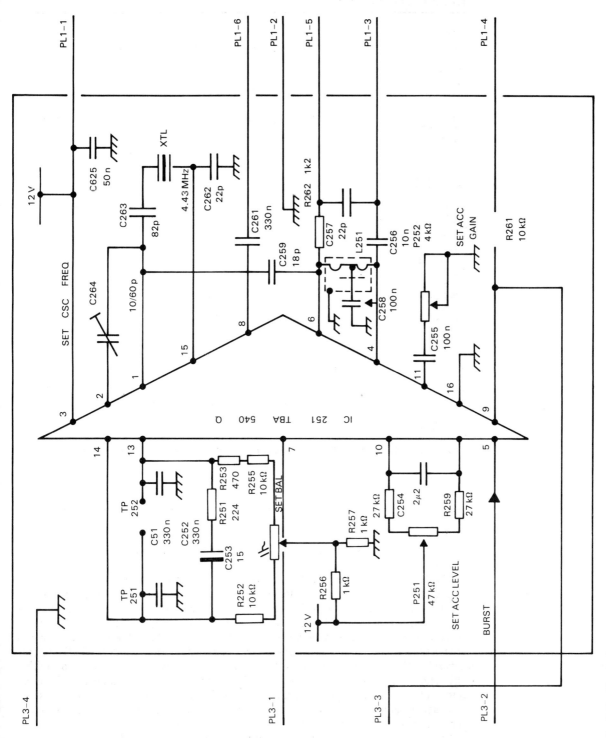

Fig. 2.7 Circuit diagram

Table 2.7   CADET input file

```
.PCB
.RES
.BOA                              The following data define PCB corner coordinates:
L5
100 100
320 100
320 300
100 300
.LIB                              The following data call component information from the library:
.REM RESISTORS
L30 28 6 2
3 3 5
25 3 5
.REM CAPACITORS
L21 20 9 2
3 5 5
17 5 5
L22 18 4 2
3 2 5
15 2 5
L23 30 13 2
4 6 5
6 6 5
L24 35 10 2
3 5 5
32 5 5
.
.
.
.EOD                             End of component data

.PCB
.CON                             The following data define circuit connections:
.COD 2
IC1 3 C14 2 PL1 1
C14 1 PL2 4
C13 1 IC1 1 C12 1 C9 1       Pin 1 of capacitor 13 to pin 1 of integrated circuit 1, etc.
.
.
.
.EOD                             End of connection data.
```

between components. Fig. 2.9 shows the PCB after completion of initial component placement.

Step 3.   The connections between components must now be converted to suitable 'routes' for copper tracks. To preserve circuit function these tracks must not touch one another unless they are intended to be connected. Thus, tracks which must cross have to do so on different layers of metallization. In this example both sides of the board are being used as surfaces on which copper tracks will be etched. The designer converts connections to routes using the interactive editing facilities. Modern CAD systems generally offer facilities for automatic placement and automatic routing of PCBs, although for densely packed designs it is often necessary to provide a degree of manual intervention. For example an auto-router might achieve typically 90% success in converting connections to routes on a complex PCB. The designer completes the remainder working interactively.

There are often other important considerations which lead to certain parts of a PCB

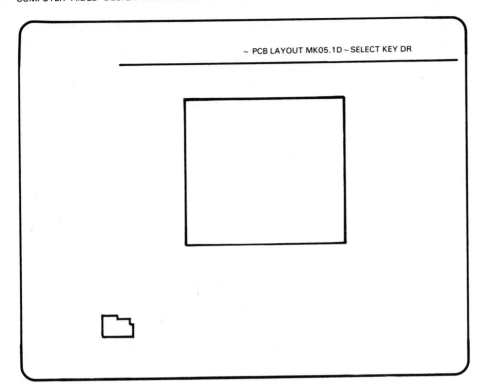

Fig. 2.8   Display prior to placement

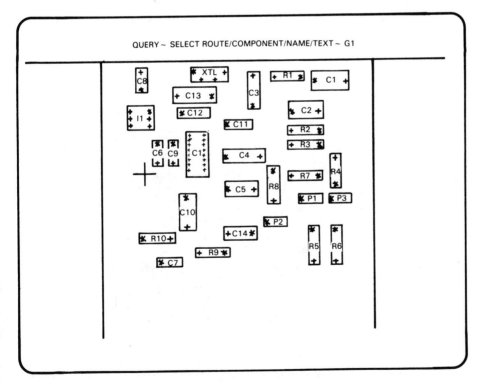

Fig. 2.9   Placement completed

being routed manually — for example where special requirements of track proximity or characteristic impedance would take longer to describe to the CAD system than to enter manually.

The designer carries out the process of placement and routing using a range of commands which in this case are selected via the keyboard. The CADET system provides several powerful editing commands including:

CORNER	which inserts a corner in a track at a position determined by the stylus moving on its tablet.
SWAP SEGMENT	which allows a track to be moved to the other surface of the PCB in order to cross another track.
MOVE	which permits components to be moved from their initial positions.
ROTATE	which swings a component through 90% each time the stylus tip is depressed on the position corresponding to the particular component.

Step 4. When the designer is satisfied with the outcome of a command sequence, he signifies this to the CAD system via the CONFIRM key. Ultimately the layout is completed (Fig. 2.10) and ready for checking. Apart from verifying that the connections between components are all present, checks should be made that design rules have not been violated. These include, for example, spacings between tracks and between adjacent pads. In this example the checking software has been run and has verified that there are no design-rule violations.

Step 5. The results of this work can now be stored for future use, or transmitted

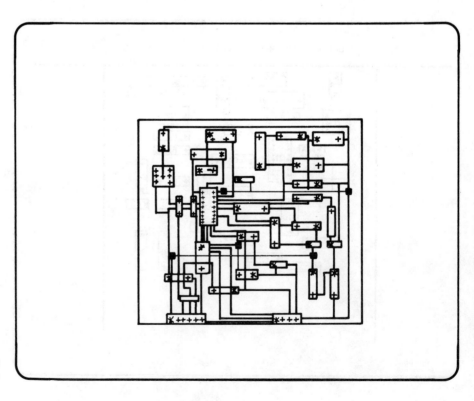

Fig. 2.10   Completed PCB design

directly to the production area in the form of data for the plotting of artworks; solder resist and silk screening patterns; and other machining and assembly operations.

2.9.3 Surface-mounted assemblies and hybrid microcircuits

The above description of PCB design by computer is directly applicable to both SMA and hybrid forms of circuit realization. The construction of hybrid microcircuits was discussed in Chapter 1, where it was noted that more than one layer of metallization is often employed. Surface mounting assemblies are increasingly being produced using traditional PCB substrate materials, and several layers of metallization are often necessary in these types of applications.

Thermal simulation of PCBs is becoming more important as packing densities increase with the adoption of surface mounting technology. Recent developments in this area include CAD software which can be employed for thermal analysis both in the initial conceptual and in the final stages of PCB design[28].

Where operation over a wide temperature range is envisaged, laminates with controlled thermal expansion coefficients are required in order to minimize the mechanical stresses set up at the joints between components and substrate. These stresses may be reduced in the 'steady state' by close matching of the thermal expansion coefficients of substrate and components. Unfortunately, temperature-coefficient matching does not totally solve the problem. Stresses are induced at switch-on due to the much shorter time constants of the components — the source of heat dissipation — compared with the time constant of the substrate itself. Selection of the correct substrate for an SMA may involve an initial attempt at layout and routing prior to finally confirming the number of interconnection layers, and hence of selecting a suitable substrate material.

Providing the very fine lines and generally tighter packing densities can be accommodated, a conventional PCB CAD system can be used to design hybrid microcircuits and surface mounting assemblies (Fig. 2.11). It will, however, be necessary to create a new component library to include suitable surface mounting versions of more traditionally packaged components.

Whilst providing improved packing density, and hence potential cost reductions due to reduced enclosure costs and perhaps reduced board count, SMA technology offers even greater benefits when we consider automatic assembly of components onto substrates. However, it must be noted at the design stage that access to the connections of each component from underneath the board are not generally available, and this will have significant implications in respect of testing of the finished assembly. The problems of designing for testability are therefore in some respects similar — for a hybrid microcircuit in particular — to those of silicon integrated circuits. This subject is considered further in Chapter 7.

III — Silicon compilation

2.10 INTEGRATION OF CONCEPTUAL AND PHYSICAL DESIGN

The first two sections of this chapter focused attention upon conceptual and physical aspects of design respectively. The protracted timescale for conversion of a circuit into a

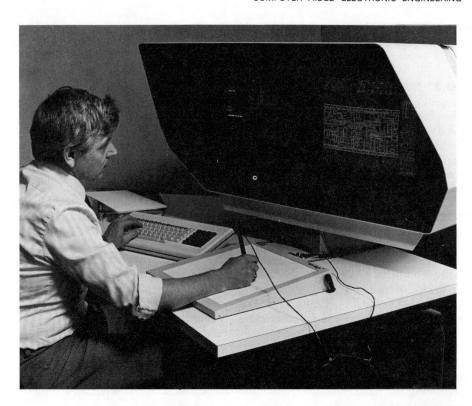

Fig. 2.11   CAD of a thick film hybrid (Courtesy *The Plessey Co Ltd)*

completed physical layout has perhaps been mainly responsible for the separation of these two aspects of design, which are often undertaken by different members of the design team. There are often disadvantages in the separation of these tasks. For example a circuit designer may have to wait for several days to obtain a layout of a PCB which, with just a minor change in the circuit, could have been routed satisfactorily with one less layer of metallization, perhaps obviating the need for a multi-layer board.

There is no fundamental reason why these two stages of design should be carried out entirely separately or even sequentially since, whilst a layout cannot be completed until the circuit is fully defined, global layout and provisional routing of bus systems can provide valuable feedback to the designer as the detailed circuit definition proceeds in a hierarchical 'top-down' approach. Recently the development of new and more powerful software tools has enabled this philosophy to be put into practice in the area of silicon design.

2.10.1   Hardware description languages

Software systems which convert a high-level description of a logic system to code which can be understood by the subsequent design and verification systems — such as logic simulators and mask-making equipment — are known as hardware description languages (HDLs). One purpose of an HDL is to avoid the need to represent a design in diagrammatic form, since with increasing scale of integration this form of representation becomes impracticable.

Prior to the development of HDLs, designs had to be described using very low-level languages such as Caltech Intermediate Format[29], an interchange format developed to provide a means of representing shapes corresponding to the various integrated circuit mask layers in a format which was independent of the particular machine chosen for mask- making.

Higher-level structured languages were developed[30] as circuit complexity increased, and the silicon compiler is a logical development of this work aimed at providing a process-independent description of how the circuit should function, together with information on how the required function could be implemented.

2.10.2 The universal silicon compiler

Realization of the engineer's dream, that it would one day be possible to describe a requirement in natural language to a computer capable of processing the requirement and of controlling a manufacturing facility to produce a silicon circuit capable of satisfying the requirement, still seems a long way off. A necessary step towards this idea of a 'silicon synthesizer' is a design automation tool known as a silicon compiler[31].

2.10.3 Current silicon compilers

A silicon compiler is similar in concept to a compiler for a high-level language such as, for example, a Pascal compiler, which creates machine code upon which a microprocessor can operate directly.

Whilst currently still in an embryonic stage of development, the silicon compiler is a significant response to the needs of system designers who require ASICs but cannot justify dedicating the majority of their time solely to acquiring the skills of driving the many individual CAE/CAD tools which have previously been necessary for the development of custom integrated circuits. There is much controversy as to what, precisely, constitutes a silicon compiler and in particular whether a CAE workstation with intercommunicating packages for hierarchical description, logic simulation and verification, test-pattern generation and physical design is or is not a silicon compiler. Rather than enter this discussion, we will concentrate here upon the practical implications of using a silicon compiler for the most popular type of semi-custom integrated circuit, the gate array.

Truly universal silicon compilers do not yet exist, but several systems are currently operational[32], most of which were initially launched as systems dedicated to one particular device technology, such as CMOS or NMOS, and are now progressively being improved towards the ultimate goal of universality of technology.

2.11 LATTICE LOGIC CHIPSMITH™ COMPILER

As with any advanced software tool, the user of a silicon compiler has not so much bought (or bought time on) a fully developed product as access to a continually developing capability. The system described below, CHIPSMITH from Lattice Logic Ltd, was originally devised as a compiler for CMOS gate arrays. CHIPSMITH has been further developed with facilities for designing standard-cell circuits as well as optimized arrays in a range of manufacturing technologies.

Versions of this system have been supplied for running on a wide range of mainframe and minicomputers and upon engineering workstations. The output from CHIPSMITH consists of mask-making data (in a range of optional formats) and, via the

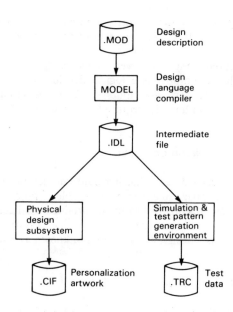

*Fig. 2.12 CHIPSMITH
process flow (Courtesy Lattice
Logic Ltd)*

in-built simulator, test tapes to run on industry standard integrated circuit test machines.

The design process flow for CHIPSMITH is shown in Fig. 2.12. Input to CHIPSMITH is via its structured, high-level hardware description language, MODEL. A logic system is described in hierarchical form, with each block definition relating eventually, via progressively lower layers, to the level of pre-defined primitive parts contained in the system libraries. Interconnections between parts are by means of named signals or vectors of signals (buses). The user may enter the MODEL description directly or with the assistance of a schematics editor, DRAFTSMITH™, which is part of the Lattice Logic suite of CAE programs for ASIC design. DRAFTSMITH converts a schematic diagram built up on the CAD screen into a MODEL description.

Once the hardware description is completed, a compiled version of MODEL is created and stored in an Intermediate Design Language (IDL) file. The IDL file is similar in many respects to a 'net list' used in PCB design systems and is the input to the simulator and also to the physical design system.

The simulator in the CHIPSMITH suite is called SWITCHSMITH™, an event-driven MOS simulator operating at switch level. It handles the circuit states 0, 1, and x (unknown). Electrical parameters, including delay times, are taken from files relating to the particular target process in which the chip is to be fabricated. Thus the simulator ensures that both functional and timing requirements are met prior to physical design. Later in the design cycle when mask patterns have been prepared, actual wiring capacitances and resistances can be fed back into SWITCHSMITH and a re-simulation executed. This enables the simulator to make a more accurate prediction of the timing properties of the circuit under simulation than would be possible from its topological properties alone. In this way the system helps to ensure that layout delays will not cause timing-related problems.

SWITCHSMITH can also be used as a fault simulator. 'Stuck at' faults can be applied to the circuit in order to evaluate the fault coverage of a particular set of test vectors. Once satisfactory fault coverage has been demonstrated, the results can be combined with the circuit description to create (automatically) a chip test program for running on the Fairchild 'SENTRY' product tester.

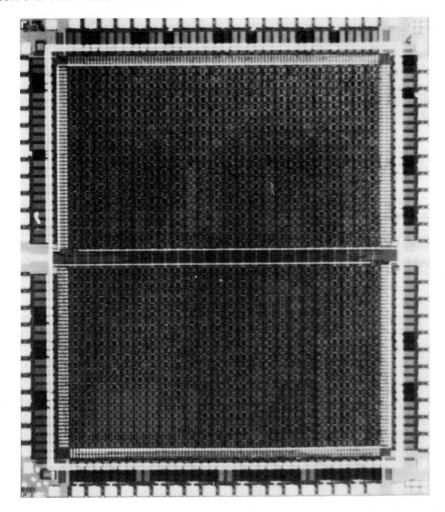

Fig.2.13 Gate array chip layout (Courtesy Siliconix)

The physical design process begins with declaration of the chip image for the base array. The number of columns, rows per column, channels, stages per row, etc, are defined. Thereafter, personalization of the gate array is entirely automatic. The placement algorithm breaks the design down hierarchically to produce a notional one-dimensional array of transistor primitives called from the basic parts library, and arranges them so as to minimize the amount of wiring between parts. The one-dimensional array is then split into several pieces which are superimposed upon the surface of the chip image. (Lattice Logic have also developed a two-dimensional placement algorithm to complement the one-dimensional system described above.) The auto-router then wires the separate parts together via the wiring channels.

The final stage is mask preparation. The output from the physical design system is fed (e.g. in Caltech Intermediate Format) to the mask preparation software. Data compatible with a range of optical and electron beam masking machines can be obtained by selection of the appropriate conversion software.

Systems such as CHIPSMITH can help OEMs take advantage of the move towards 'silicon foundries' (where the semiconductor manufacturer's involvement is limited to provision of silicon processing capacity), since it is relatively easy to change from one fabrication process to another. The parameters of the new process must be fed into the physical design system and the simulation repeated. It is even practicable to move a commercially successful chip from gate array technology to standard cell realization once volume sales are assured, since the physical design system can produce the necessary mask preparation tapes automatically. Thus an OEM who uses this type of facility is at worst only one processing step away from having a second source should he meet insuperable supply difficulties via his chosen fabrication process.

SUMMARY

At circuit-level the primary aids to design are analysis programs. Using accurate models of the active and passive circuit elements the performance of new circuit structures can be forecast with confidence. At the higher functional levels, hardware description languages are used to simulate functional blocks of circuitry. The techniques of hierarchical design allow either a 'top-down' or a 'bottom-up' approach to be applied. Thus a system can be realized by working outwards if necessary from practical constraints of what could or should be built upon a single silicon chip or printed circuit assembly.

CAD techniques for physical design must take into account not only the constraints and design rules of the medium of realization, but also the need for the finished product to be testable. Simulators can provide the means of establishing a set of test patterns to obtain a high percentage fault coverage against 'stuck at' faults. In practice this approach provides for effective testing of most types of logic circuits.

Although PCBs and hybrid microcircuits are sometimes purchased as standard 'components' (e.g. single-board computers), the majority are fully custom built. Silicon design using full custom integration is justified on cost grounds only for very large-volume products. Semi-custom integration, using CAD techniques, permits cost-effective use of ASICs in a wide range of OEM products.

REFERENCES

1. Van Valkenburg, M.E. (1960) *Introduction to Network Synthesis*. Wiley, New York.
2. Belevitch, V. (1958) Recent developments in filter theory. *IRE Trans.*, **CT-5**, 236–52.
3. Wolfendale, E. (Ed.) *Computer Aided Design Techniques*. Butterworths, London.
4. Gummel, H.K. and Poon, H.C. (1970) An integral charge control model for bipolar transistors. *Bell Syst. Tech J.*, **49**.
5. Ebers, J.J. and Moll, J.L. (1954) Large signal behaviour of junction transistors. *Proc. IRE.*, **42**, 761–72.
6. Cohen, E., Vladimirescu, A. and Pederson, D.O. (1979) *User's Guide for SPICE*. Version 2E.1B. University of California, College of Engineering.
7. O'Reilly, W.P. (1975) Transmitter power amplifier design — Part II. *Wireless World* **82** (1478).
8. The 'SIMON' MOS/CMOS simulator from ECAD Inc is an example of a fast simulator with dependable convergence. This simulator can accept SPICE-coded network files.
9. Russell, G. Kinniment, D.J. Chester, E.G. and McLauchlan, M.R. (1985) *CAD for VLSI*. Van Nostrand Reinhold (UK), Wokingham.
10. Lewin, D. (1977) *Computer Aided Design of Digital Systems*. Crane Russak, New York.
11. Ambler, T. and Musgrave, G. (1985) Test pattern generation techniques. *Silicon Design* **2** (4,5).

12. Bennets, R.G. (1984) *Design of Testable Logic Circuits*. Addison-Wesley, Reading, Mass.

13. Roth, J.P. (1966) Diagnosis of automata failures: a calculus and a method. *IBM J. Res. Dev.*, **10**, 278–91

14. Bennets, R.G. (1981) *Introduction to Digital Board Testing*. Crane Russak: New York/Edward Arnold: London.

15. Goldstein, L.H. (1979) Controllability/observability analysis of digital circuits. *IEEE Trans. Ccts and Syst.* **CAS26** (9), 85–93.

16. Wadsack, R.L. (1978) Fault modelling and logic simulation of CMOS and MOS integrated circuits. *Bell Syst. Tech. J.* **57**, 1449–74.

17. Smith, C.O. (1976) *Introduction to Reliability in Design*. McGraw-Hill–Kogakusha, New York.

18. Dawson, P.M. (1985) A perspective for the '80s. *New Electronics*, 8th Jan 1985.

19. Goodwin, M. (1985) Semi-custom LSI - reducing costs and lead times. *Electronics Industry*, **11** (8).

20. Bastian, J.D. *et al.* (1983) Symbolic parasitic extractor for circuit simulation (SPECS). *Proc. 20th Design Automation Conf.*, 346–52.

21. Alexander, W. (1985) MOS and CMOS arrays. In *Gate Arrays — Design and Applications*. Read, J.W. (Ed). Collins, London.

22. Hurst, S.L. (1982) Commercial applications of semi-custom IC design. *Proc 2nd Int. Conf. Semi-Custom ICs*. Prodex Ltd, London.

23. Irissou, P. (1982) Combining linear and digital functions on single bipolar/CMOS/ULA chips. *Proc 2nd Int. Conf. Semi-Custom ICs*. Prodex Ltd, London.

24. Lee, C.Y. (1961) An algorithm for path connections and its applications. *IRE Trans. Electron. Comp.* **EC10**(3), 346–64.

25. Aramki, I. *et al.* (1971) Automation of etch pattern layout. *CACM*, **14**, 720–30.

26. Kelly, J. (1985) Placement and routing techniques for gate arrays. *New Electronics*, 47–50 (March)

27. Kernigham, B.W. Schweikert, D.G. and Persky (1973). An optimum channel routing algorithm for polycell layouts of integrated circuits. *Proc. 10th Design Automation Conf.*, 50–59 San Diego, California.

28. Kumar, C. (1985) Thermal simulation of printed circuit boards. *Electronic Product Design*, 39–42 (June)

29. Mead, C. and Conway, L. (1980) *Introduction to VLSI Systems*. Addison-Wesley, Reading, Mass.

30. Breuer, M. (Ed.) (1982) *Computer Hardware Description* Languages and their Applications. North-Holland, Amsterdam.

31. Ayres, R.F. (1983) *Silicon Compilation and the Art of Automatic Microchip Design*. Prentice-Hall, Englewood Cliffs, New Jersey,

32. Gray, J.P. *et al.* (1982) Designing gate arrays using a silicon compiler. *Proc. 19th Design Automation Conf.*, Las Vegas.

3 CAD equipment

3.1 OBJECTIVES

This chapter focuses upon the hardware systems which are used in CAD, with particular emphasis on equipment suitable for electronic engineering applications. These systems are the engines which enable a designer to access the CAD techniques discussed in the previous chapter.

The chapter is divided into two parts. In part I, following a brief outline of the evolution of CAE, the main elements of a CAD system are discussed. Part II consists of descriptions and comparisons of various types of CAD equipment currently in use for electronic design engineering.

In reading this chapter you should obtain an insight into:

— the evolution of electronic computers
— the development of interactive computer graphics
— common forms of graphic input and output devices
— microcomputer CAD systems
— 'turnkey' CAD systems
— engineering workstations
— CAD bureaux

I — Elements of CAD systems

3.2 EVOLUTION OF THE ELECTRONIC COMPUTER

The history of calculating machines can be traced back a very long way in time. For example, it is recorded that the ancient Greeks devised automata using water and steam power to execute sequences of actions. We need to move on to the seventeenth century, however, for evidence of the first machines being devised to handle mathematical tasks automatically.

3.2.1 Early calculating machines

The first mechanical calculating machine[1] was built by Wilhelm Schickard in 1623, and could add and subtract numbers automatically. It is Charles Babbage, however, who is generally held to be the 'father of modern computing'. Babbage developed two important machines in the early 1820s: a 'difference engine' and an 'analytical engine' as they were termed. Although these machines never achieved a properly functional state, they contained all the key elements of a modern computer: a means of INPUT, a means of STORAGE to hold numbers and instructions during computation, an ARITHMETIC UNIT in which calculations were performed, a CONTROL UNIT and a means of OUTPUT for results. Babbage's task was made all the more difficult by his insistence on using decimal arithmetic. Had he been aware of the mechanical simplicity of binary arithmetic, Babbage's machines might well have fulfilled the ambitions of their creator.

The first electrical computing machine, devised by Herman Hollerith (1860–1929), was used to count and to help sort coded cards. The cards had twelve rows of holes the positions of which signified personal information such as sex, religion, etc. This machine was used in the American census of 1890 and was the forerunner of punched card machines still used today in certain applications for input and output of computer data. Electro-mechanical technology, in the form of banks of relays, was the basis of automatic calculators developed in the USA by Howard Aiken and George Stibitz at the end of the 1930s.

3.2.2 Early electronic calculators

The concept of a universal machine, capable of solving a variety of mathematical problems, was first propounded by the British mathematician Alan Turing[2] in 1936. (Turing also produced the first chess-playing computer program, and contributed to early work into artificial intelligence.) During World War II Turing worked for British Intelligence using Colossus 1, a machine purpose built and solely dedicated to helping decipher Enigma, the German communications code system. In 1944 at the University of Pennsylvania, J Presper Eckert and John W Mauchly completed ENIAC (Electronic Numerical Indicator And Calculator) which was essentially a general purpose electronic calculator containing over 18 000 vacuum valves[3]. The program was fed into ENIAC via a plug board, which meant that changing the program was a time-consuming task. John von Neumann extended the work of Eckert and Mauchly, building EDVAC (Electronic Discrete Variable Automatic Computer) between 1945 and 1949[4]. This time the machine actually stored the program in memory.

3.2.3 The first commercial computers

In 1950 the first commercial machines appeared on the market. These first-generation computers included the Manchester University/Ferranti Mark 1 in the UK, followed soon after by the UNIVAC 1 in the USA. Early commercial computers used vacuum valves and mercury delay-line memories, and as a consequence they suffered from slow speed and poor reliability.

Second-generation computers, using transistors instead of valves, began to appear in the late 1950s, ten years after the discovery of the bipolar transistor. One of the earliest solid-state computers, the Atlas Guidance Computer, was used from 1958 to control satellite launching at Cape Canaveral.

Silicon integrated circuits provided a further reduction in physical size and resulted, in the 1970s, in third-generation machines which, while far from portable, were at least transportable.

3.2.4   Present-day computers

The current generation of computers ranges from large-scale scientific 'supercomputers' to compact portable business and personal microcomputers. From but a few tens of hand-built machines in 1950, computer technology has penetrated most aspects of industrial, commercial and domestic life.

The characteristics of the major categories of present-day computers — potentially the 'engines of CAD systems' — are outlined below. In Section 3.4 the way in which some of these computers are currently being applied to CAD is discussed.

Supercomputers

There is currently a limited market, perhaps just a few hundred machines in total, for very powerful 'number-crunching' supercomputers. Cray Research, established in 1972 by Seymour R. Cray, dominates this niche of the market, producing machines capable of performing many hundreds of millions of floating-point operations per second (megaflops)[5]. This is the sort of power which is needed, for example, to process the data collected by numerous sensors worldwide in order to produce, within one or two hours from the time of data input, accurate and up to date weather forecasts. These types of machines had a typical selling price of around £10 million in 1985.

Mainframe computers

These powerful machines are generally used in large electronic data processing (EDP) applications such as banking and insurance, and in large company accounts, payroll and general administration. Their performance, while dwarfed by that of a supercomputer, is still measured in tens of millions of instructions per second (MIPS). IBM dominate the mainframe market. Their top of range machines include the 3090/200, launched in 1985, with 64 Mbytes of main memory and providing 26 MIPS at a cost at launch date of £4.3m.

Minicomputers

Originally, minicomputers were built with 16-bit word length (occasionally 24 bits), an early example being the DEC PDP7 from Digital Electronic Corporation. In recent years 'superminicomputers' having 32-bit word length and offering greater speed and multi-user operation have formed the basis of a number of successful Computer Aided Design (CAD) systems. Digital remain market leaders in this area with a large number of CAD systems suppliers having adopted the DEC 'VAX' range of machines. Other leading supermini suppliers include Prime and Hewlett Packard, both of whom supply complete CAD/CAM systems. A typical mid-range machine cost £40 000 in 1985.

Microcomputers

The development of the microprocessor — the central processing unit of a computer on one silicon chip — has led to the availability of low-cost personal and small business computers. Initially these machines employed 8-bit word length and were severly limited in the amount of memory which they could rapidly address[6]. (An 8-bit machine can directly address $2^8 \times 2^8$ words, or 'bytes' of memory, i.e. 65 536 bytes, which in computer terminology is referred to as 64 Kbytes.) In 1979 these types of machines came onto the market at prices of a few hundred pounds. The Zilog Z80 and Intel 6502 chips are examples of microprocessors which proved popular for such systems.

The Motorola 68000, a 16-bit microprocessor, has been adopted in a large number of scientifically orientated microcomputers since its launch in 1979/80. IBM chose the Intel 8086 as the basis for their Personal Computer. Launched in 1981, the PC has been adopted as a 'standard' for business and, to a lesser extent, scientific applications. Numerous software houses have produced applications programs for IBM PC and compatible machines. The more powerful IBM XT and AT models have found favour as computing 'engines' in several CAD and CAE systems, examples of which are discussed later in this chapter. These more powerful microcomputers can address large amounts of random access memory (RAM). A typical installation costing around £3500 in 1985 would include 0.5 Mbyte RAM and a sealed magnetic (Winchester) hard-disk store providing back-up memory capacity of ten Mbytes or more in addition to 'floppy' disks capable of holding up to one Mbyte.

Supermicros

The availability of true 32-bit microprocessors, such as the National Semiconductors NSC32032 and the Motorola 68020, has led to a new generation of 'supermicrocomputers' which are compact and transportable but yet offer performance capable of competing in roles traditionally filled by minicomputers. These types of computers represent the fastest growing sector in the CAE processor market.

3.3 COMPUTER GRAPHICS

Many of the early scientific computers were, of course, applied to the solving of equations in engineering design. CAD really began to play a major role in product innovation with the introduction of visual means of man–machine interaction — when computers were first able to display visual images which could be modified by a user interacting with the computer via some sort of input device.

3.3.1 Evolution of computer graphics

Among the earliest computers with a visual display unit (VDU) as an output medium was 'Whirlwind I', a machine produced at the Massachusetts Institute of Technology (MIT) in 1950[7]. Whirlwind was able to generate pictures upon a cathode ray tube (CRT) display. However, a decade or more passed before significant interest was shown in computer graphics. In 1962 Sutherland's 'Sketchpad'[8] demonstrated dramatically that a person could interact with a computer in a relatively fast (compared with defining features by sets of mathematical coordinates) and convenient way. From this work a whole new industry evolved in the 1960s and 1970s, with graphical representation of information taking over from textual descriptions almost entirely in many areas of design. Tektronix has been the pioneering market leader in graphics display terminals, providing the display systems for several CAD vendors' products. Recently Tektronix have broadened their product offer by the acquisition of CAE SYSTEMS Ltd, and now supply complete CAD systems.

The development of powerful computer graphics systems has enabled the design of the complex semiconductor devices essential to the continued advancement of computer graphics technology.

3.3.2   Current graphics display technology

The CRT remains by far the most popular display device at the present time, although it now has serious competition from liquid crystal displays (LCD) and plasma panels. The most commonly used types of graphics displays are discussed below, beginning with the two types of CRT displays which have been widely adopted for current CAD systems:

Vector scan

The picture on a CRT is produced by selective illumination of phosphorescent regions (called *pixels*) on the screen. One system of display uses vector (or stroke) writing techniques (Fig. 3.1). By driving both vertical and horizontal beam deflection systems the spot can be made to trace a straight line at any angle across the tube screen. To avoid visible flicker effects, the image may be re-drawn upon the screen at a rate of 25 frames per second or greater (this technique is variously referred to as *direct-beam refresh, vector-refresh* or *stroke-refresh*).

Raster scan

In most modern systems an electron beam is made to scan the face of the tube in a series of horizontal lines called a *raster* (Fig. 3.2). Picture information is held in an electronic screen memory, or *frame buffer*, which normally resides inside the display terminal, and is passed to the CRT drive circuits in order to modulate the intensity of the electron beam and hence to illuminate the appropriate pixels.

In a monochrome display the frame buffer needs only to store signals of logic levels 1 or 0 corresponding to black and to white pixels on the raster. Thus a single bit of frame buffer can store the status of one pixel. To modify the picture the user needs some method of altering the contents of the frame buffer, either as a result of running a pro-

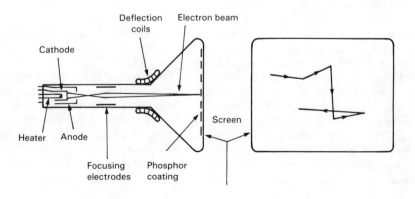

Fig. 3.1   Vector scan CRT display

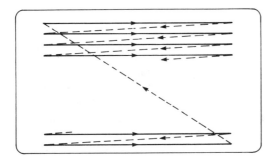

Fig. 3.2 Raster scan display

gram or of feeding in data to the computer. Raster scan graphics cannot perfectly reproduce straight lines, circles etc., due to the quantization of the image into discrete pixels. The severity of 'staircase' effect depends upon the definition of the graphics system, which in effect is determined by the resolution of the CRT and the size of the frame buffer. High-resolution display monitor tubes are much more complex, and hence more expensive, than tubes used for 625 line television.

A state-of-the-art high-resolution display of 4096 × 4096 pixels requires a screen memory capacity of 16 megabits (2 megabytes) in monochrome or four times that figure (2^4 = 16) for 16-colour display if a full screen of pixel data is to be stored. CRT displays in current electronic engineering CAD systems typically provide a resolution of 1024 × 900 pixels, but higher definition is often available for systems which are also used for three-dimensional solid and surface modelling[9]. Normally a limited number (eg 4, 8, 16, . . . 256) of colours can be displayed at one time; these being selected from a 'palette' often of a much greater number of colours. Some graphics terminals have several 'pages' of screen memory and offer the facility of opening 'windows' in the displayed screen (the 'current' screen) to look through to details in other pages of display memory. These windowing facilities play a key role in modern CAE workstations, and are discussed further in Section 3.10.

Direct-view storage tube

A type of CRT graphics display which was used extensively during the 1970s and early 1980s involved the use of a direct view storage tube (DVST)[10]. A DVST display retains an image drawn by the electron beam, and hence avoids the need for re-drawing (termed *refreshing*) the screen display unless changes are made to the image.

Liquid crystal displays

While limited in colour range, speed of response and feature definition, LCD panels require far less electrical power, are more compact and should in the near future be producible to high-performance specifications at costs well below those of CRT technology[11]. The LCD has another limitation, however: it does not emit light, and so it provides a clearly readable image only over a limited range of background lighting conditions. LCD displays do, however, provide excellent visibility in bright sunlight. Viewing angle is restricted compared with most other types of displays, but this is rarely a serious limitation when used as a fixed computer display.

Plasma panels

Plasma panel displays are another flat-screen technology which is finding applications in computer graphics systems. These devices rely on the ionization of a gas, which emits light, when a high voltage is applied between two electrodes. The electrodes are immersed within the gas. In d.c. plasma displays the electrodes actually contact the gas, while a.c. displays involve a thin layer of insulation between the electrodes and the gas. (The latter type create less electrical interference and, manufacturers claim, offer longer lifetime than d.c. types.) The red–orange light output of a plasma panel is of higher brightness than a CRT and permits an exceptionally wide viewing angle — up to 160° for a plasma panel compared with typically 70° for a CRT. Plasma panels offering resolution up to 1728 × 1280 are currently available.

Other display technologies

Several other techniques, including electroluminescent panels, are emerging to challenge the traditional CRT display for computer graphics output[12]. To date none of these technologies can match the resolution of the CRT, and it is reported[12] that market surveys into CRT demand indicate a likely growth of around 100% between 1985 and 1992.

3.4 PLOTTERS

The traditional design process culminated in a set of drawings and manufacturing instructions defining the components, sub-assemblies and completed equipment to be manufactured. At some stage in the future it may well be that 'designers' will simply input a statement of requirements and wait for a production prototype to be built, with little or no human operator intervention. For the present, however, there is a demand for equipment to convert a visual image on the display of a CAD system to hard copy of a quality as good as if not better than the displayed image. Colour displays predominate in this area of CAD, and they certainly do ease the task of visualization.

The demand for colour hard copy is increasing correspondingly, and this need is currently addressed by several types of plotters. The following is a brief investigation of their characteristics:

3.4.1 Pen plotters[13]

These represent the most mature plotting technology, and microprocessor-based machines with a high degree of intelligence, capable of plotting circles, filling areas in colour, scaling, etc., automatically on receipt of simple commands from the host computer, have been in common use since the late 1970s. Such features relieve the main processor load since data, transmitted in a high-level language, can be stored in a buffer within the plotter while the host computer continues with other tasks.

In addition to lines, text is generally required on an engineering drawing, and most plotters contain firmware (software procedures stored in a permanent form such as a non-volatile Read Only Memory, or ROM) so that a variety of character fonts can be produced, often at any required size and angle. Pen plotters can be categorized according to how the pens and paper are made to move relative to each other:

Flat-bed plotter

Here the paper remains stationary while either continuous servo-motors or stepper motors drive the pen in two orthogonal directions. A 'pen-lift' mechanism is actuated when it is necessary to move from one region of the drawing to another without producing a line. Lines are produced by the plotter as follows:

(a) Select pen of the required colour.
(b) Move to x and y coordinates corresponding to start of line.
(c) Place pen down onto drawing surface.
(d) Create line, using appropriate firmware routine (e.g. interpolating between data points, circle function, etc.).
(e) Lift pen from drawing surface.

This procedure may appear quite obvious, but it contrasts with the method of driving other types of plotters, as discussed below.

Drum plotter

Here the pen movement is restricted to the horizontal direction, while vertical movement is obtained by rotating a drum carrying the paper. This has the advantage that although the width of paper is constrained by drum dimensions, there is in principle virtually no limit to the length of plot which can be produced. On some drum plotters, pens are stored at the side of the plotting surface, so the pen carriage must move to the side for each pen change. An alternative to this approach is where a set of pens is situated together on the carriage and the appropriate colour pen is put down onto the drawing medium upon command.

Moving-bed plotter[14]

A hybrid of the above techniques provides a plotter having a flat bed which moves in one direction (e.g. vertical) while a pen carriage moves in the other (horizontal) direction. The paper is securely gripped between a pair of grit-covered wheels and the plotting bed.

3.4.2 Pen plotter media and pen types

The types of drawing medium selected depends upon application. For simple check plots low-cost paper suffices, but if good dimensional stability is required, precision mylar film sheet or roll is used, while clear plastic film is needed for overhead projection. Appropriate pens must be used for each medium and many modern plotters are capable of accepting roller ball, felt tip, pencil or ink-feed draughting pens. Several of these machines have the ability to detect the type of pen which the user has fitted to the carousel or pen carriage, and to adjust pen speed automatically to obtain best results. In general a slower pen speed is required when working with ink-feed pens compared for example, with pencil.

3.4.3   Photoplotters

Where plotting onto a light-sensitive medium is required, as for example in the creation of artwork for PCBs, a 'light head' can be fitted to some types of flat-bed plotters. Laser heads are used when the very highest resolution is required and these can use scanning techniques (similar to the electrostatic plotters discussed below) to provide speed advantages over non-scanning photoplotters. Recent developments include flat-bed plotters capable of accepting interchangeable optical and ink pen heads.

3.4.4   Electrostatic plotter[15]

Whether the computer display is in vector or raster scan, data to be plotted must be fed to an electrostatic plotting head in the latter form, since this type of plotter uses a fixed writing head with a large number of fine styli which selectively place electrostatic charge on the paper at positions corresponding to the data. Liquid toner is applied via a fountain, particles of which adhere to the charged areas of the paper. For colour plotting, four toners are required: Magenta, cyan, yellow and black. Other shades are obtained by overlaying the three coloured toners in a multiple-pass method. The paper must be rewound to the start position before each new colour is added to the plot in order to achieve registration between colour overlays. Special electrostatic media must be used with these types of plotters, and are available as translucent or opaque paper or stable plastic film.

Plots consist of a series of dots which are overlapped by up to 50% to form a high quality line. The resolution of an electrostatic plotter is determined by the number of dots per mm. Current machines provide typically 15–25 dots/mm. For 'draught' outputs it is sometimes possible to select a lower resolution and hence increase plotting speed. The speed of an electrostatic plotter, in contrast with pen plotters, is independent of the density of lines in the plot. Thus for complex drawings, maps, etc., this type of machine greatly out-performs the pen plotter in terms of speed. The running costs of electrostatic plotters are, however, rather greater than many other types of hard copy units.

3.5   OTHER HARD-COPY OUTPUT DEVICES[16]

For high-quality text output, required for user handbooks etc., daisywheel printers are most commonly used. Their ability to reproduce graphics is limited to simple black-and-white line diagrams made up of the fixed character set of the daisywheel. For reproduction of complex graphics, dot matrix printers can be used. These are available in two basic types:

3.5.1   Impact printers

These reproduce characters and graphic symbols by means of a line or an array of print hammers which strike an ink ribbon in contact with the paper. A typical 'near letter quality' printer has a row of 18 or 24 needle hammers. Speed of printing is increased by such features as bi-directional printing and logic-seeking. (The head carriage does not travel along a line scan any further than the last item of data to be printed on that line.) Dot matrix impact printers using three ink ribbons are available for reproduction of colour graphics and, while they cannot compete in terms of quality with a pen plotter, they offer speed advantages especially when used in 'draught' mode for check plots.

3.5.2 Ink-jet printers

Ink-jet printers employ a stream of tiny droplets of ink formed via a fine nozzle and then deflected either onto or away from the paper under the control of the plotting data. Again the three colours, magenta, cyan and yellow are overlayed as required to build up a wide range of ink colours. Early ink-jet systems suffered from reliability problems due to clogging of the jets, but in later models this problem appears to have been overcome by the incorporation of automatic jet-capping and self-cleaning facilities.

3.5.3 Thermal printers

Colour hard copy can also be obtained by means of thermal printing. Pigment is transferred from a donor sheet, carrying ink-impregnated wax, by means of a line of fine wire nibs in the thermal printing head. These nibs are heated up instantaneously to create the required dot patterns. A triple-pass technique is employed for each of the colours magenta, cyan, yellow, and like electrostatic plotters the system must rewind to a reference position in order to achieve registration of the colour overlays.

3.5.4 Laser printers

When monochrome reproduction is sufficient, the very high speed and quality of a laser printer may provide accurate drawings with the added advantage of virtually silent operation.

3.6 COMPARISON OF PLOTTER CHARACTERISTICS

In Table 3.1 some of the main characteristics of various types of plotters are compared. It should be stressed that the prices of all types of plotters are continuing to fall as improvements in design result in reductions in the number of mechanical parts and in simplified assembly. Self-test and fault diagnosis facilities are built in to some of the more advanced machines. The price ranges quoted for a particular type of machine reflect not only the variations in performance specifications but also the availability of

Table 3.1

Plotter	Type	Paper size	Resolution (mm)	Speed (mm/s)	Typical price range* (£K)
Hewlett-Packard 7475	Moving-bed 6 pen	A4/A3	0.25	400	0.5–3
Nicolet Zeta 824/826	Drum 8 pen	A4–A1 & roll	0.025	650	3–15
Versatec ECP-42	Electrostatic 256 area colours	To A0 & roll	0.125	27 000 mm^2/s	20–50

* Price ranges refer to plotter types, not to specific examples quoted.

optional features such as large buffer memory to free the host computer during protracted plotting routines.

Plotting speed comparison

Because of the differing plotting processes it is not possible to make direct speed comparisons between, for example, pen plotters and electrostatic plotters, since the plotting time of the former is proportional to total line length while for the latter time is proportional to plot area.

For a typical 300 mm × 200 mm PCB artwork produced at four times full size, a pen plot of one set of tracks was completed in twenty minutes. The same plot was completed in less than thirty seconds using an electrostatic plotter. The speed ratio becomes much greater when very complex plots, such as mask patterns for vlsi chips, are considered. For this reason electrostatic plotters, despite their high capital and running costs, are used extensively in such applications.

3.7   GRAPHICS INPUT DEVICES

Early batch processing machines were fed with inputs from punched (Hollerith) cards or paper tape, which were the normal means of creating and retaining a permanent copy of programs or data. Interactive computing requires a much more flexible input method. Graphics commands can be fed into the computer via an alphanumeric keyboard, but when large amounts of data or commands are required this too is a very slow and error prone method. Various input devices are in common use in modern CAE systems, and some of them are briefly described below:

3.7.1   Graphics tablets and digitizers

In the early 1960s, Thomas Ellis of the Rand Corporation devised the Rand Tablet, a flat insulating plate in which are embedded a number of parallel horizontal wires and a similar number of parallel vertical wires. The wires carry coded signals which can be picked up and amplified by a sensitive probe or stylus. A computer input is connected to the stylus which is moved on the rectangular surface such that its position is interpreted by the computer as a moving dot or cross (cursor) on a CRT. When a switch is operated, by depressing the stylus tip, a trace is recorded on the screen.

Reserved areas on the tablet can be selected to convey particular commands to the computer. For example to change line width and colour or to select functions for drawing rectangles and circles. Complex procedures such as copying a block of pixels, zooming to magnify a feature or erasing part of the screen picture may also be called up via such an input device. Various magnetic, optical, acoustic and electrical methods are also used in graphics tablets and digitizers (these are large, high-precision tablets) to provide coordinate information in digital form. Acoustic systems of three orthogonally positioned microphones are used to create the equivalent of a three-dimensional tablet for input to certain solid modelling CAD systems.

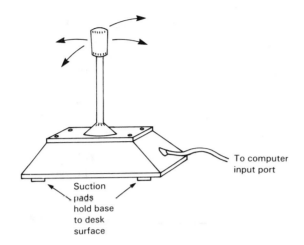

Suction
pads
hold base
to desk
surface

To computer
input port

Fig. 3.3   Joystick

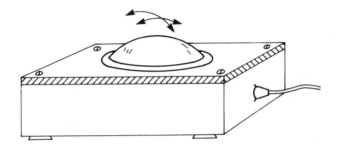

Fig. 3.4   Tracker ball

3.7.2   Joystick and tracker ball

Other methods of obtaining a moving cursor are via a joystick (Fig. 3.3) or a tracker ball (Fig. 3.4). Since the user's hand stays in one region, his eyes can concentrate on the screen picture. These are less costly devices than the graphics tablets, but do not offer the same degree of linearity between hand movement and cursor position.

3.7.3   Mouse

The mouse is a small plastic device moved around the desk on two wheels which are set at right angles to one another. Each wheel drives an optical shaft encoder which sends a series of pulses to the computer as the wheel rotates. Thus a cursor on the screen can be made to move in unison with the motion of the mouse. Push buttons on the top of the mouse allow commands to be fed to the computer.

For example, with suitable software, drawing functions such as zoom, copy, etc., can be selected by moving the cursor to a region of the screen which is reserved for the required function (Fig. 3.5). A button situated on top of the mouse is then pressed. When the cursor is next moved to a region on the screen, marked for example by a rectangular frame, all features within that region are processed by the selected function.

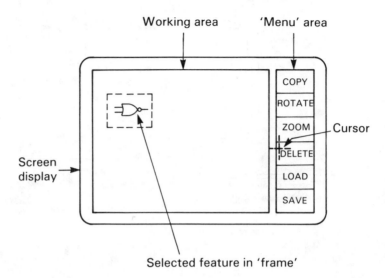

Fig. 3.5 Selection of commands by cursor and menu

3.7.4 Light pen

A light pen consists of a light-sensitive probe which can be pointed at a region of the screen. From knowledge of the time during the frame scan at which the small group of pixels pointed to are illuminated, the computer is made to calculate the corresponding screen coordinates. The pen itself consists either of a photocell and amplifier linked to a bistable circuit, or a length of fibre optics cable connected to a photocell at the computer input. The light pen can be used for selecting commands from a 'menu' in much the same way as described for the mouse.

II — Current CAD systems

3.8 CAD ON MICROCOMPUTERS

Although microcomputer-based CAD systems as simple replacements for the conventional drawing board were developed in the late 1970s, it was the introduction of 16-bit machines which led to the popularity of personal computers as vehicles for serious CAD and CAE work.

3.8.1 Eight-bit systems

Initially, 8-bit microcomputers, with their generally modest computational and graphics performance, were applied as 'erasable sketchpads' with the ability to produce tidy drawings of moderate complexity. The simplest systems use a drawing cursor controlled from the keyboard which, while very slow, can be made quite accurate by the provision of panning and zooming facilities. Thus the user can effectively look at an enlarged view of just a small section of his drawing.

Commands are provided for regularly used functions such as the drawing of arcs, circles and rectangles and line drawing aids. For example a 'rubber banding' facility maintains a straight line between the current cursor position and the previous stored cursor position. The line can be altered in length and orientation by moving the current cursor position, and when the user is satisfied with its location he can freeze the line by a confirmatory key command.

Disk-based microcomputer systems of this type generally have facilities for creating and storing often-used features, such as component outlines, which can be recalled and 'dragged' across the screen to the desired location. Such a facility speeds diagram preparation and leads to improved consistency in drawings. Trackerballs, mice or graphics tablets are available as improved input devices for serious CAD use, and output hardcopy is provided by means of a printer and/or plotter.

Dot matrix printers have been made to create usable PCB artwork masters at 2:1 scale by means of a multi-pass technique. Pen plotters are commonly used for artwork generation and, in cases where only moderate definition is required, simple systems are available which plot with a single line representing the copper track of the PCB. In these systems the line width must be altered by multiple pen changes in order to obtain conductor tracks of various widths. More sophisticated systems draw multiple parallel overlapped lines to achieve the effect of varying track widths.

3.8.2 Sixteen-bit systems

Microcomputer-based CAD systems dedicated to artwork preparation provide facilities for creating and storing libraries of data relating to component shapes and pin connections. Similarly, a library of standard PCB profiles can be established and called up when a new design is being started.

CAD software packages to run on 16-bit personal microcomputers are most suited to schematics data-capture and PCB and semi-custom SIC layout, and may have some limited facilities for logic simulation and verification, and for autorouting. The more powerful systems may include some facilities for checking electrical and physical design rules.

In order to illustrate typical capabilities of current CAD systems it is necessary to discuss specific examples. In this and in the remaining sections of this chapter some typical working systems are described in some detail, together with a brief survey of some other systems in the same categories.

3.8.3 IBM PC Range with Racal–Redac REDLOG/REDBOARD

Racal–Redac provide a complete PCB design facility based upon IBM PC hardware. This microcomputer-based CAD system (Fig. 3.6) can be used as a stand-alone station or confifured as part of larger system operating in a CIEE (Computer Integrated Electronic Engineering) environment, sharing a common database with other stations based upon PCs, Apollo workstations or DEC VAX computers.

REDLOG is a software package for the capture of circuit or logic data in schematic form. It supports a true hierarchical design approach and comes complete with a library of widely used electrical, electronic and logic symbols and a library of industry standard components such as TTL and CMOS devices. The libraries may be extended or modified to meet particular requirements. The output from REDLOG is a net list at gate level. This net list, which is in ASCII format, may be transferred to a logic simulator or to a PCB layout package.

REDBOARD provides automatic routines as well as interactive manual facilities for

Fig. 3.6

component placement and routing of connections. An optional mouse is available as the user interface. The system can be driven with manually created data, or it can be fed automatically from REDLOG. REDBOARD comes complete with a library of physical data on devices in common use, and extension of the library is straightforward. Full dimensional checking routines are available and post-processing permits outputs to be fed to either a pen plotter or a photoplotter.

An example of a REDBOARD screen display is illustrated in Fig. 3.7. A valuable feature of this system is the provision of full-width track display. Output from a pen plotter can, if required, be used directly for the creation of prototype PCBs.

3.8.4   Other microcomputer-based CAD systems

A range of microcomputer-based CAD systems are outlined in Table A3.1 in the apendix to this chapter. It should be noted that this and subsequent tables in the

Fig. 3.7 REDBOARD screen display

Appendix are not intended as an exhaustive survey, but are provided only as a representative sample of the sorts of CAD systems currently available.

3.9 DEDICATED 'TURNKEY' CAD SYSTEMS

The major suppliers of electronics CAD and CAE systems have since around the mid-1960s progressively developed a range of computer-based tools specially for electronics design. These were completely self-contained hardware and software systems with which the purchaser simply 'turns the key' to get the system running. Initially these turnkey vendors concentrated on the task of PCB design, but latterly most turnkey vendors have broadened their scope to cover the demand for vlsi design aids for ASICS.

3.9.1 Low-cost entry systems

At the lower end of this market, single-user machines suitable for interactive layout of single or double-sided PCBs provide outputs on cartridge tape. The tapes can then be sent to one of the many 'post-processing' bureaux which convert the data to photographic artworks using a photoplotter. Initial data input to these smaller machines is often in the form of a 'net list', or circuit description (e.g., Pin 4 of IC12 connects to pin 6 of IC15, etc), which can be prepared either on the CAD station or, to improve throughput, on a compatible low-cost microcomputer as only a text file is involved. With the inclusion of optional hard disk facilities, some of these low-cost entry CAD systems are capable of handling multi-layer boards and provide limited auto-routing and design rule checking of the completed layout (checking of actual wiring connections versus the net list and also track and pad clearances).

One disadvantage of some of these simpler CAD stations is the display form which they use. The conductor routes are all shown as lines of a single width, and it is necessary to create a final copy plot in order to see the tracks as they will appear on the actual PCB. This causes few problems when an auto-router is used, but if routes have to be inserted interactively it is very easy to violate the physical design layout rules and to have tracks touching one another or touching other pads. Displays at true line-width are a valuable feature for this type of work, since it is very difficult to spot errors of this type when looking at the display of a large and complex PCB. Of course, the error-checking software normally highlights such problems at final check, but substantial rework may then be necessary.

3.9.2 Multi-user CAD systems

For users having a higher board design throughput, and perhaps demanding greater capability in terms of complexity, mainframe- or minicomputer-based systems capable of accommodating a number of simultaneous users (typically three to five design stations) are available from the leading CAD system suppliers. The majority of these up-market systems can be used for very much more than just the design of PCBs. For example with the right software installed they can be used for vlsi chip design and for mechanical design and draughting.

For PCB work a full set of manufacturing outputs is provided. Normally this includes:

— schematic diagram
— components list
— artworks for each layer of the board
— data for numerically controlled drilling
— data for numerically controlled profiling
— silk screening artwork
— solder masking artwork.

The data from the CAD system can also be processed for driving automatic or semi-automatic assembly equipments and automatic test equipment.

3.9.3 CALMA multi-user systems

Typical features of a modern multi-user CAD system are described below by reference to CALMA, now part of the General Electric Corporation, who have held a dominant position in the electronics CAD system market — most notably in the area of silicon integrated circuit design systems — for more than a decade. CALMA offer a range of single and multi-user systems and were one of the first CAD system suppliers to develop a coherent policy for integration and customer growth routes by upwards compatibility.

Figure 3.8 illustrates the CALMA philosophy to CAD/CAE system integration in which any number of 'streamlined' Eclipse 'Central Support Units' (CSUs) and work-stations based on 32-bit CPUS may be linked together via a local area network. The

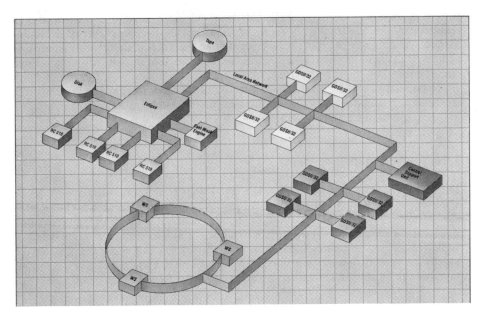

Fig. 3.8

GDSII/32 workstations are sufficiently powerful to perform all foreground tasks, such as PCB or ASIC design, via their local processors, and only need to access the CSUs for background data.

A wide range of line, electrostatic and photoplotters is supported and the CARDS hierarchical PCB design software suite is available via a database management system which recognizes not only the PCB graphics but also the textual library information necessary for design and manufacture of the PCBs.

CALMA multi-user systems are based upon the Data General Eclipse S140 computer with floating-point unit for rapid processing of double-precision (16 decimal digit) floating-point arithmetic. The systems can support four workstation terminals which provide rapid user response since the operating system can manage files and peripherals in background mode.

CALMA also offer PCB design solutions on Apollo computers using their T-Boards product.

3.9.4 Other turnkey systems

In Table A3.2, in the Appendix to this chapter, the key features of several other turnkey systems currently in use are summarized.

3.10 ENGINEERING WORKSTATIONS

Whereas turnkey CAD systems have evolved via the computer-aided design draughting route, engineering workstations have developed in response to the needs of the conceptual design stage. Thus electronic engineering workstations initially concentrated upon the mathematical aspects of design engineering — simulation and test pattern generation — but they have now been enhanced to address the total design requirements of electronics systems engineering[17]. Facilities are available for hierarchical design of the functional concept of an electronic system; for its simulation and modelling; and its physical design in PCB, SMA, hybird microcircuit or integrated circuit form.

The popularity of the workstation as a design medium is underlined by the growth in sales of major equipment vendors. Both Daisy Systems and Mentor reported an increase in turnover of around 300% during 1984.

3.10.1 Workstation concept

The general concept of a workstation is a stand-alone equipment based upon a 16- or 32-bit processor. Rather like a powerful personal computer, a workstation has a large main memory — at least 512K — but, unlike the former, it has a large, high-resolution monochrome or colour screen with bit-mapped graphics, and a multiple-window facility[18]. Thus, for example, it is possible to display a circuit in one window and the waveforms from its simulation in another window. The size and position of the various windows can be altered as required by means of menu-accessed commands.

3.10.2 Networking

Although each user has a powerful computer contained within his workstation, a key feature of these tools is their networking capability by means of which users can share data via a high-speed data network. This is a particularly effective way of providing complete integration of the various CAE processes provided that suitable software, in the form of a database management system, is available.

With manufacturers of superminicomputers adding intelligence and improved graphics capability to the terminals of their multi-user systems, such facilities offer many of the features normally associated with workstations.

3.10.3 Software portability and UNIX

While suppliers of supermini-based turnkey systems have in the past provided a limited range of their own proprietary software to run on their machines, several workstation vendors have sought to achieve compatibility with a very wide range of CAE software. The obvious benefit to the owner of such a system is that he can purchase or lease whichever software solutions best suit his needs, and as he introduces the new software it is not necessary to learn a new machine interface. Popular logic simulators, such as GenRad's HILO-2, for example, are available on several types of workstations. There is a danger, however, in this approach that software systems from a range of suppliers may prove more difficult to integrate.

Originally conceived as a program development environment, the first UNIX system[19] became operational in 1971. Initially, the UNIX operating system found favour with universities and research establishments, but latterly many CAD/CAE workstation vendors have recognized the benefits of ease of portability of software which UNIX affords, and are now offering a UNIX environment with their products.

Popular processor chips for workstation applications include the 16-bit Intel 80286, Motorola 68000 family, and the 32-bit NSC32032 from National Semiconductors.

3.10.4 Daisy workstations

Following its inception in the Autumn of 1980, Daisy Systems Corporation achieved a growth rate in sales, during the first four years of its existence, of around 300% annually as its CAE workstation concept received rapid acceptance, first by the silicon design industry and more recently by a wider range of scientists and engineers seeking a powerful but flexible high-productivity design environment. While Daisy also produce CAD/CAE systems based upon personal computers, here one of Daisy's more powerful 'one system per engineer' workstations will be described in some detail. (A range of Daisy systems is available covering all aspects of CAE from concept to silicon mask data or PCB layout data[25].)

Daisy MegaLOGICIAN/PMX.

The MegaLOGICIAN workstation is based upon the Intel iPAX286 microprocessor family which has 16-Mbyte memory addressing capacity, on-chip virtual-memory support, and a floating-point accelerator (Intel 80287) for computation-intensive tasks such as running SPICE analyses[20]. The workstation hardware is shown in Fig. 3.9, and the internal system architecture is illustrated in Fig. 3.10. A key feature is the use of several processors, each selected for optimum performance in its particular role (e.g. I/O management, high speed graphics, etc.). This workstation can take a logic concept from design entry through simulation and verification.

A major problem in preparing data for simulation of large systems is the construction of accurate models of complex physical parts. Where these parts already exist in hardware form it seems rather unfortunate that time has to be spent in creating a software description in order to simulate the interaction between the physical part and the rest of the system under development. PMX — Physical Modelling Extension — is the name given to Daisy's solution to this problem. Virtually any hardware device, be it a microprocessor, gate array, printed circuit assembly or whatever, can be connected directly into the system simulation. In practice the accuracy of simulation is greatly improved by this technique, which effectively offers the benefits of speed and thoroughness of logic simulation with the added assurance of virtual breadboarding at the highest functional hardware level for existing parts of the system.

The simulator (MDLS), with its accelerator, provides a truly interactive design medium even for complex logic circuits containing hundreds of thousands of gates. For example, a 1200 100-ns clocks simulation on a 1500-element CPU chip has been shown to take just three seconds. The simulator supports mixed-modes and provides functional and timing verification in addition to the physical modelling extension facilities. Fault simulation facilities are enhanced by a testability analyser (DTA). If the circuit turns out to be hard to test the DTA can suggest various alternative solutions to enhance the circuit's testability.

Multiple screen windows

A feature which is particularly useful when debugging complex logic systems is the ability to call up windows onto the screen to display waveforms at selected circuit nodes while the schematic is also visible.

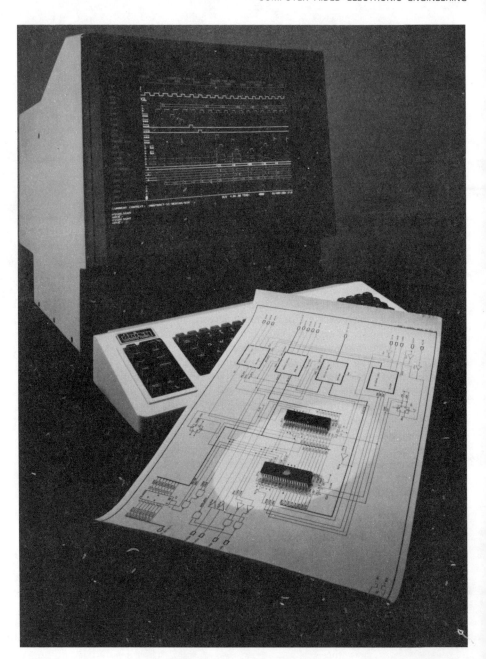

*Fig. 3.9 Daisy
MegaLOGICIAN workstation*

Design example

This example illustrates in graphical form the capabilities of the MegaLOGICIAN and
PMX system specifically in the verification area. The circuit is based upon the Intel
8086 microprocessor and the top-level block diagram is shown in Fig. 3.11.

(a) *Drawing creation.* The logic schematics for the CPU, the Memory I/O, and the

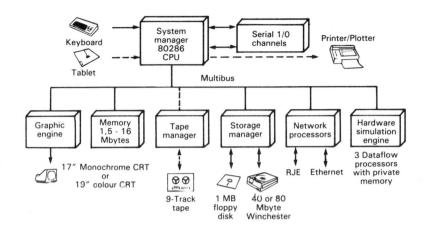

Fig. 3.10 MegaLOGICIAN system architecture

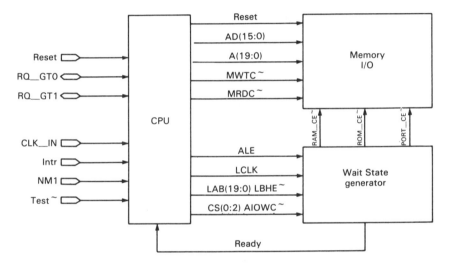

Fig. 3.11 System block diagram

Wait State Generator are illustrated in Figs 3.12, 3.13 and 3.14. Each of these drawings was created using the MegaLOGICIAN's Drawing Editor. Whilst working in the Drawing Editor it is possible to move rapidly up or down the hierarchy (e.g. from block diagram to schematics) or across the design tree (e.g. from CPU schematic to Memory I/O schematic).

(b) *Design rule checking.* Once the design has been entered via the drawing editor, the pages are run through the MegaLOGICIAN's compiler and linker where they also undergo extensive electrical and design rule checking.

(c) *Data definition.* Before starting the simulation the user defines both the stimulus to be applied to the circuit and the initial data contents of registers and memories.

(d) *Simulation.* The entire circuit, including the 8086 CPU (for which, because a software behavioural model was not available, the PMX was used), was simulated for 125 000 time units, corresponding to 1000 clocks of the 8086. The total run time was less than one minute. Mixed-mode simulation was used: PMX for the CPU; functional level for the ROM, PAL, and MSI circuits; and gate-level for the test of the random logic.

(e) *Waveforms.* Running the logic simulator produces waveforms like those of

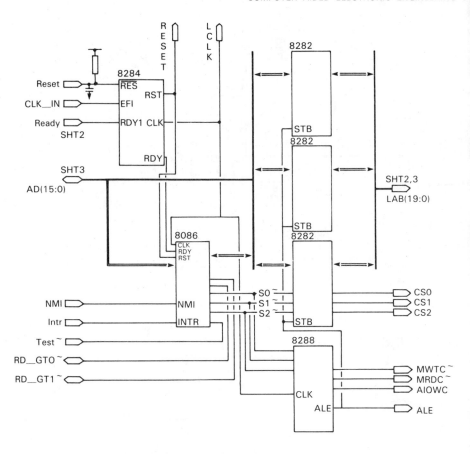

Fig. 3.12   CPU outline schematic

Fig. 3.15. To generate the exact waveform display required, the user calls up the schematic, selects the signals to trace from the drawing and issues a PROBE command. Once selected, the signals can be activated as needed (displaying the strength of the ABUS signal, for example) by the FORMAT command, as in Fig. 3.16. Alternatively, the signals displayed can be selected with the FORMAT command alone, without PROBE.

(f) *Debugging*. During the debug phase, the captured data from the simulation can be viewed with Daisy's Virtual Logic Analyser. For example, the occurrence of a signal or a combination of signals is defined in a TRIGGER file (Fig. 3.17). When the specified sequence of events on those signals is met the display is halted. In this example the TRIGGER file illustrates how to command the system to find the first memory write cycle after the access to memory position OOFF (ABUS = $OOFF_{HEX}$), then to search for the fetch of the next instruction from memory (Status lines = 4), and finally to centre the display around this condition (CENTRE command).

In this example, the ABUS represents a latched address bus and the ADBUS is the data bus. Tracing the 8086's activity, in the first machine cycle (T1) the ALE goes high and the status lines are presented. Notice that the Ready line is activated to create a wait state. Data (F800) is written to address OOFFF while memory write, MWTC~, is active low. In this bus cycle the 8086 is performing a memory write function (Status line = 6) and the RAM_CE~ is active. In the following bus cycle, the 8086 does an instruction fetch (Status lines = 4). This time the data is fetched from the program ROM, as

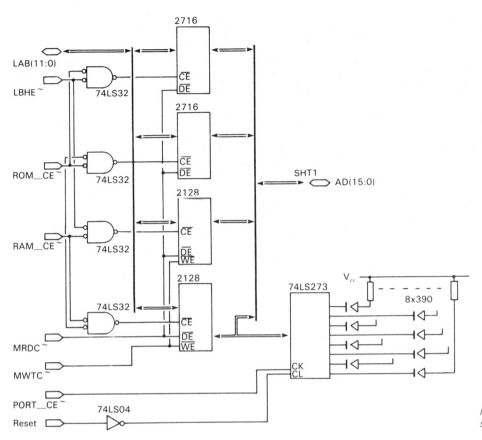

Fig. 3.13 Memory I/O outline schematic

indicated by the ROM chip enable (ROM__CE˜), and memory read (MRDC˜) is active. The final bus cycle illustrated in the waveforms of Fig. 3.15 is a memory read (Status lines = 5).

3.10.5 Other CAD/CAE workstations

Table A3.3 in the Appendix to this chapter outlines characteristics of several workstations which are currently used in electronic engineering applications. A definitive list of all such facilities would soon become outdated in this fast-moving area of technology. The interested reader is referred to specialist publications[21] and to journals which regularly survey the CAD/CAE market[22,23].

3.11 CAD BUREAUX

Several small electronics companies — and even some quite large ones — find that, either for financial reasons or because their need for such services fluctuates greatly, it is not practicable to establish their own CAD facilities. Instead they choose to sub-contract this type of work to one of the many specialist companies which have been set up to

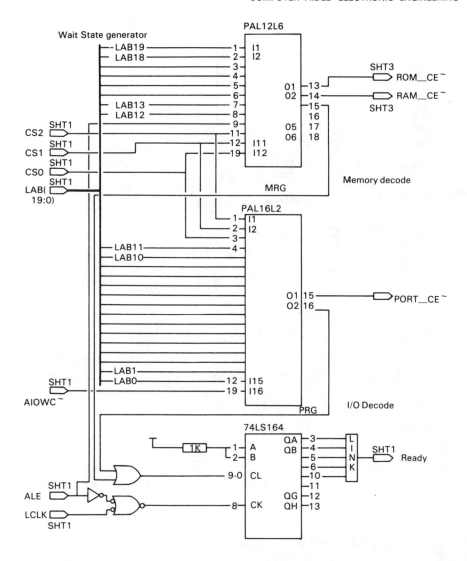

Fig. 3.14 Wait state generator schematic

provide CAD consultancy services. Some of these agencies can provide a wide range of services, including some or all of the services described below.

3.11.1 PCB post-processing bureaux

These organizations will accept the output (e.g. in tape cassette form) from a PCB design station and will use their in-house pen plotters and photoplotters to create precision manufacturing artworks, assembly overlays and other manufacturing data. Costs are usually quoted on a 'per unit board area' basis. For urgent tasks, twenty-four hour turn-round can generally be provided when necessary.

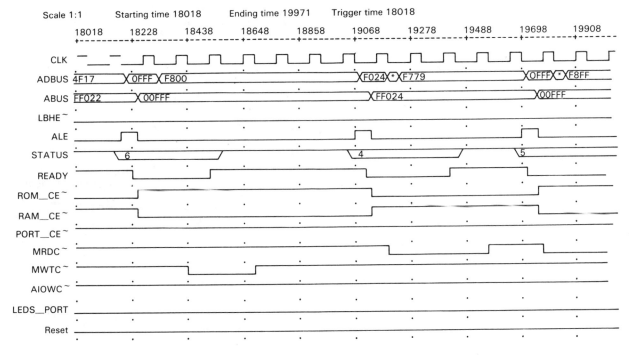

Fig. 3.15   Waveforms

```
** REPLACE **

NAME        BASE    POLARITY STRN    TRC?    SIGNAL_LIST


CLK         BIN     +        OFF     ON      LCLK
ADBUS       HEX     +        OFF     ON      AD15,AD14,AD13,AD12,AD11...AD0
ABUS        HEX     +        OFF     ON      LAB19,LAB18,LAB17,LAB16...LAB0
LBHE~       BIN     +        OFF     ON      LBHE~
ALE         BIN     +        OFF     ON      ALE
STATUS      HEX     +        OFF     ON      S2~,S1~,S0~
READY       BIN     +        OFF     ON      READY
ROM_CE~     BIN     +        OFF     ON      2:ROM_CE~
RAM_CE~     BIN     +        OFF     ON      2:RAM_CE~
PORT_CE~    BIN     +        OFF     ON      2:PORT_CE~
MRDC~       BIN     +        OFF     ON      MRDC~
MWTC~       BIN     +        OFF     ON      MWTC~
AIOWC~      BIN     +        OFF     ON      AIOWC~
LEDS_PORT   HEX     +        OFF     ON      3:XSIG67,XSIG64,XSIG66,XSIG63,
                                             XSIG65,XSIG62,XSIG60,XSIG61
NMT         BIN     +        OFF     OFF     NMI
INT         BIN     +        OFF     OFF     INTR
RESET       BIN     +        OFF     OFF     RESET
```

Fig. 3.16   Format file

3.11.2   PCB design bureaux

Design departments rarely have a steady work-load throughout the year, and design consultancies offer a service to deal with the peaks in demand. It is possible to sub-contract specific tasks to these bureaux, such as the design of a PCB, or in some cases to rent time on their CAD systems. This latter facility is particularly useful when evaluating different CAD systems prior to purchasing, or for initial training of staff

NAME	OCCUR	INTERVAL	CLK	ADBUS	ABUS	LBHE⁻	ALE	STATUS	READY	ROM_CE⁻	RAM_CE⁻
BASE			BIN	HEX	HEX	BIN	BIN	HEX	BIN	BIN	BIN
POLARITY			+	+	+	+	+	+	+	+	+
STRENGTH											
RESTART											
OFF											
TRIGGER											
FIND	1		X	X	XXXX	00FFF	X	X	6	X	X	X
THEN	1		X	X	XXXX	XXXXX	X	X	4	X	X	X
CENTER	1		X	X	XXXX	XXXXX	X	X	4	X	X	X

Fig. 3.17   Trigger file

whilst awaiting delivery and installation of a new CAD system. By this means staff can build up the necessary libraries of component and board data and so may already be at or near to maximum efficiency on the new system from the first day of on-site operation.

3.11.3   Silicon brokerage

For an OEM who is unlikely to require more than an occasional custom or semi-custom ASIC, specialist silicon design companies exist. They can provide assistance in design and procurement of an integrated circuit or, if required, they will undertake the whole task from device definition to supply of production quantities. These specialist organizations — some within established semiconductor fabrication companies, but an increasing number quite independent of the chip manufacturers — use their own CAD/CAE facilities, producing outputs in various formats via a range of post-processors which are capable of driving a range of mask-making machines. Thus the end-user is not necessarily tied to one silicon supplier, as the silicon design bureau can place the actual fabrication work with the most appropriate 'silicon foundry'. The silicon design bureau normally accept bare wafers from the processing sub-contractor and manage the assembly, testing and coding of the chips for the OEM.

Some of these companies are prepared to rent their design facilities, or to provide access to them through data terminals via which means the OEM can develop his own expertise in silicon design.

An example of such a scheme is FALCON, a service introduced by Micro Circuit Engineering to provide OEMs with low-cost prototypes of gate array chips. To do this MCE produce personalization masks which allow up to 50 different circuits to be proto-typed on a single wafer. Designs of up to 1100 gates can be processed and ten samples of each prototype are produced and supplied for testing by the OEM. Final production quantities can be left in gate array technology or the circuit can be further optimized if and when volume of sales justifies a change. For moderate volume requirements, this type of services is likely to attract growing interest in the future[24].

SUMMARY

The advent of solid state devices, and in particular integrated circuits, has provided the technology for powerful computers suitable for tackling a wide range of engineering tasks. The microcomputers of today have a performance equalling that of minicomputers of only a few years ago.

The ability to work interactively with a computer, using an input device and a graphical display, has been the major factor responsible for the increasing use of CAE techniques.

CAD facilities are available on a range of computers. Microcomputer-based systems vary from simple drawing board replacements to complete PCB design systems including many design automation features.

Turnkey systems are specially designed for CAD. They may be purpose-built machines, or built up by the system supplier from equipment and software items from a range of companies. For large design departments, multi-user installations are available offering facilities for several users to work at terminals communicating with a central computer.

Individual workstations containing their own powerful central processors can be networked so that data and peripherals are shared between a number of design stations. Features such as the ability to use physical components within what is otherwise a mathematical model of a system assist in the simulation of complex logic structures prior to their physical realization.

Hard copy output is obtainable from printers and plotters. Pen plotters provide high accuracy at moderate cost, although their speed makes them unattractive for the production of complex plots. Dot matrix techniques are used in low-cost printers which, using multiple pass techniques, can provide useful hard-copy outputs for some purposes. Electrostatic plotters and laser plotters are more expensive, but can produce high-precision complex plots very rapidly.

REFERENCES

1. Moreau, R. (1983) *The Computer Comes of Age*. The MIT Press, Cambridge, Mass.
2. Bowden, B.V. (Ed.) (1953) *Faster than Thought*. Pitman, London.
3. Stern, N. (1981) *From ENIAC to UNIVAC*. Digital Press, Bedford, Mass.
4. Pugh, E.W. (1984) *Memories that Shaped an Industry*. The MIT Press, Cambridge, Mass.
5. Capron, H.L. and Williams, B.K. (1984) *Computers and Data Processing*. 2nd Edn. Benjamin/Cummings, California.
6. Downton, A.C. (1984) *Computers and Microporcessors*. Van Nostrand Reinhold (UK), Wokingham.
7. Redmond, K.C. and Smith, T.M. (1980) *Project Whirlwind — the History of a Pioneer Computer*. Digital Press, Bedford, Mass.
8. Sutherland, I.E. (1962) *Sketchpad: a Man–Machine Graphical Communication System*. PhD Thesis. MIT.
9. Holt, M.G. (April 1985) Experiences with CAD solids modelling and its role in engineering design. *Computer-Aided Engineering Journal*, 2(2).
10. Hobbs, L.C. (Jan. 1981) Computer graphics display hardware. *IEE Computer Graphics and Applications*, 1(1), 25–39.
11. Tassell, C. (1985) Developments in display technology. *New Electronics* 18(12).
12. Update: display technology. (1985) *Electronics Industry*, 15–37 (May).
13. Harris, D. (1984) *Computer Graphics and Applications*. Chapman and Hall, London.
14. HP7475A from Hewlett-Packard is a popular A4 moving bed plotter.
15. Harrison, N. (Aug. 1984) Raster colour on paper. *Systems International*, 12(8), 34–37.
16. Pipes, A. (1985) Plotters: expensive afterthought. *CADCAM International*, 4(8).
17. Carlstedt-Duke, T. (April 1985) The CAE workstation — how much can it do?. *Computer-Aided Engineering Journal*, 2(2).
18. Williamson, J. (1985) Benefits of workstations for full custom IC design. *Silicon Design*, 2(6).
19. Gauthier, R. (1981) *Using the UNIX System*. Reston Publishing, Virginia.
20. Paseman, W.G. (1985) Data flow concepts speed simulation in CAE systems. *Computer Design* (January).

21. O'Neil, T. (1984) *The Complete CADCAM buyers' Guide*. CAD SOURCE Ltd., Alpha House, Beech Lane, Macclesfield, Cheshire SK9 2DY.
22. *CADCAM International*. Monthly, EMAP Business and Computer Publications Ltd.
23. *SYSTEMS International*. Monthly, Electrical and Electronic Press.
24. Goodwin, M. (1985) Semi-custom LSI - reducing costs and lead times. *Electronics Industry*, **11**(8).

APPENDIX A3.1

Some features of a selection of current CAD systems are compared in the following tables. It should be noted that these represent only a very small sample of the range of existing CAD systems, and are provided as an indication of current technology.

Table A3.1 Typical microcomputer-based CAD systems

Supplier	Product	Processor(s)	Features
ASA	DIAMOND	Whitechapel MG1. HP, Plessey, Pixel 80	Logic design system and PCB design system
Computamation	EDS	Apricot IBM PC/XT Zenith 159 Sirius	PCB design system. Hard copy via dot matrix printer using multi-pass technique
Daisy Systems	Personal Logician	IBM PC, XT and AT range	Design entry and compilation. Logic simulation prior to data transfer to workstation
Hytech Systems	Microdesigner	Microcomputers operating under CP/M or MS/DOS	PCB layout system plus two-dimensional draughting facility
Personal CAD Systems	PC-CAPS PC-LOGS PC-CARDS	IBM PC (640K min) range and compatibles	Hierarchical schematics capture & editing. Logic simulation. Interactive PCB layout.

Table A3.2 Typical turnkey CAD systems

Supplier	Product	Features
Racal–Redac	CIEE	PCB and ASIC design environment consisting of PCs, Apollo workstations, Racal min- and maxi- CAD systems networked to DEC VAX computers. Software includes VISULA for PCBs and hybrids as well as suites for gate array, standard cell and full custom design.
Wayne–Kerr Datum	ARTWORKER	A range of stand-alone PCB design stations for schematics capture and PCB layout. Input is via keyboard and joystick, integral with the station. Output can be fed directly to a pen plotter or stored for transfer to a photoplotter. Monochrome or colour displays are available.
Ferranti	Silicon Design System	The ULA (Uncommitted Logic Array) is the Ferranti gate array system. Design stations and software are supplied. Based upon DEC VAX 11/730 computer.

Table A3.3 Typical CAE workstations

Supplier	Product	Features
Apollo	DOMAIN range	Hardware engines adopted by many systems suppliers including Racal–Redac, CAE Systems (Tektronix) and GE Calma. Software houses such as Lattice Logic and Silvar Lisco provide suites for running on Apollo workstations. Berkeley 4.2 and System V UNIX standards available.
Mentor		Complete systems based upon Apollo workstations. Software for schematics data capture, simulation at circuit (MSPICE) and logic levels. ASIC conceptual and physical design and verification, hardware testing and documentation. Floating-point processors available.
Valid	SCALDSYSTEM	Hardware and software produced by Valid. S-32 computer based upon 68010. Stand-alone and multi-user system available, all with Ethernet networking facilities. Links to other CAD systems, e.g. for PCB layout.

4 Computer aided manufacturing techniques

4.1 OBJECTIVES

Continuing through the product life cycle of an electronic equipment, in this chapter computer aided manufacturing and assembly techniques are discussed. In Chapter 5, ways in which these techniques are applied and organized in current production systems are considered.

While a PCB loaded with its electronic components is sometimes sold as a product in its own right, it is generally necessary to provide some form of equipment housing, whether it be a standard rack assembly, a custom designed plastic case or a sealed casting with rugged characteristics. Thus in practice, electronics engineers have to work with or be knowledgeable in the disciplines of mechanical engineers, chemist and metallurgists etc. However, in this book attention is restricted to the (already very broad) range of techniques for production of the electronic assemblies themselves. In reading this chapter you should gain an insight into:

— manufacturing philosophies for various product types
— PCB manufacturing process
— computer aided and automatic assembly techniques
— systems for plating and soldering
— techniques for achieving interconnections.

4.2 MANUFACTURING PHILOSOPHY

The optimum approach to computer aided manufacturing for a particular company will very much depend upon the nature and quantities of products to be manufactured. For example it would not be expected that a factory designed for mass production of television receivers would be ideally suited to the production of satellite power supply units in batches of less than ten of a particular design.

Much is being talked about the 'factory of the future', with a totally flexible manufacturing system capable of changing product type without stopping production for the weeks or even months which can be involved at present. While the component industry (electrical and mechanical types) can already claim to have made considerable inroads into this area, truly flexible 'assembled equipment' facilities are only now beginning to be seen in the electronics industry.

The criteria for introducing CAM could be some or all of the following:

— more rapid transfer from design completion to availability of delivered product in the required volume
— reduction of 'work in progress' (and attendant effects on cash flow) through faster and more flexible manufacturing and transfer processes
— improved quality and / or reliability of delivered product
— reduction of scrap or wastage
— reduced cost of production
— improved control through better monitoring and communication
— reduction of routine or dull work for people.

At a time of high unemployment the last point may appear somewhat contentious. However, there is much evidence that poor quality workmanship is encouraged in a work environment in which people carry out repetitive tasks with no obvious link to a final and worthwhile achievement. Some motor manufacturers have dismantled their flow production lines and replaced them by 'assembly teams' which carry out a complete assembly process. As a result both production costs and quality are higher, and it is perhaps significant that it is for the luxury car market that this job-enrichment policy has generally been adopted.

Clearly, then, a different approach is needed depending upon the relative importance of several factors, including those listed above. The manufacturing engineer must consider ways in which computer assistance can be applied to the various processes involved in manufacturing electronic equipment, looking at techniques which could be appropriate for small and medium batch production as well as those more suited to very large volume manufacture.

It must be clearly stated that the full benefits of computer aided electronic engineering cannot be obtained without linking the processes of design and manufacture. The major sources of manufacturing cost are determined during the design of a product. The designer must therefore be aware of the processes and cost implications of production.

One way of ensuring that production requirements are taken into account during design is to incorporate them into the 'design rules'. This becomes more straightforward if the CAD and CAM systems work from a common database. There is, however, a caveat: innovation in design is essential. Successful products are generally a compromise between novelty and producibility. Production engineers *must* influence design from the start of a project, but it is important that the cooperation and communication is in two directions. Product designers and production system designers need to innovate in each of these areas — design and production — and this means that standards, processes and procedures need to be controlled in an environment of continual change.

Managing in a static technology would be a relatively simple if not trivial task, but such a situation no longer exists in the electronics industry. Management of a design / manufacturing business is a particularly demanding task, but the adoption of computer aided engineering — in its broadest sense — can provide a means to better management control throughout the product life cycle. In Chapter 8 this theme is pursued further, but for the remainder of the present chapter some of the important manufacturing processes involved in production of electronic equipment will be considered.

4.3 MANUFACTURE OF PCBs

The principle steps in bringing a product idea through design and manufacture were outlined in Chapter 1. In this section more details are presented as to how computer aided systems can be used to produce the PCB in readiness for assembly of its complement of electronic components.

4.3.1 Machining operations

Figure 4.1 illustrates a flow chart for the production of a double-sided, through-hole-plated PCB of the type used within the majority of electronics products. It will be noted that several of the steps involve cutting with a saw or a router, and drilling of the PCB. Initially the board goes through processing with a border of spare material to facilitate handling and it is in this region that 'tooling' holes are drilled or punched to provide

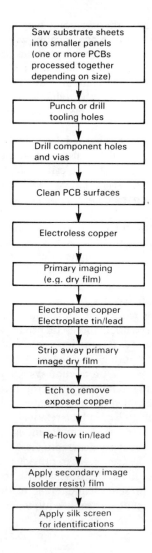

Fig. 4.1 Flow chart for PCB
         manufacture

accurately known datum points for registration throughout the subsequent processing stages. The data for relative positioning of these points comes from information defined during design of the board. Data for drilling and profiling (machining to final shape, which is often achieved using a router) is available from the CAD system used in its design.

For manufacturing departments working from manually produced designs, the drilling data can be created by means of a digitizer. A cursor is moved across a PCB layout drawing, or a blank PCB, and data are entered into a computer by pressing a button on the digitizer when the cursor position corresponds to that of a hole. Some numerically controlled machines have facilities for 'learning' the required drilling positions; thus the drill table itself acts as a digitizer.

Drilling

Numerically controlled (NC) drilling or routing machines (some machines are capable of performing both drilling and routing operations) commonly use a paper tape input: with machining parameters coded on the tape. CNC (computer numerically controlled) machines can also use paper tape input or receive input from a host computer which transfers the data to the drilling machine in electrical form. DNC (direct numerical control) machines operate under real-time control from a central computer. The computer may share its time between several machines involved in different manufacturing processes.

The majority of PCB materials are reinforced with glass fibre, which is highly abrasive. Tungsten carbide tools are generally used (in preference to steel-alloy tools which would wear very rapidly) for drilling operations, whilst the increased cost of impregnated diamond tooling is justified only for sawing operations such as roughing out blanks from large sheets of laminate. The quality of the internal surface finish is important when through-hole plating is to be used, and this in turn means that the drill speed and rate of feed must be accurately controlled and matched to the type of laminate and size of hole[1]. The machine tools used must be capable of high spindle speeds (up to 80 000 rpm when drilling 0.5 mm holes) and must be adequately guarded to protect personnel in the event of a drill shattering.

Single or multi-spindle NC machines are available, the more advanced of which not only move the tool to the required position, but also control spindle speed and feed rate, changing the tool without operator intervention after a predetermined number of drilling operations (usually about 5000 board holes between re-grinds of a drill). Thus damage due to blunt tools is avoided.

It is common practice to pin up to about four boards together in a stack to increase productivity. A backing layer is also used so that the tool does not come into contact with the metal of the machine bed. In some applications an additional (copper-free) layer is mounted above the boards to minimize burring as the drill enters the top board of the stack. A typical five-spindle NC drilling machine can produce 2000 PCB holes per minute.

Routing

The operation of routing the board profile removes its handling border and may also be used to cut slots or large holes. While the spindle of a router rotates upon the same axis as that of a drilling machine, the cutting action is quite different. The router tool moves in a horizontal direction (except when initially piercing the laminate) with the side of the tool accomplishing the cutting operation. Tungsten carbide burr-router tools carry multiple-spiral fluting in contra-rotation to create a 'criss-cross' effect of small teeth. Again it is common to stack three or four boards together if the slots to be routed are sufficiently wide to allow a large diameter (2.5–3 mm) routing tool to be used. Feed rate and spindle speed must once again be related to tool diameter in order to limit the radial load on the tool. Heavy-duty routers are constructed with special spindle bearings to cope with these stresses.

The data relating to the profile of the PCB is, of course, defined at the commencement of its design, and most CAD systems incorporate facilities for post-processing of the PCB design data into a form suitable for feeding NC drilling and routing machines.

4.3.2   Plating and etching processes

Referring back to the board manufacturing flow chart (Fig. 4.1), the drilled boards are cleaned in a brushing machine fed with a suitable cleaning solution (or simply clean water) to ensure that surfaces are free from grease etc. Electroless copper is deposited from solution to cover both the board surface and also the insides of the holes in order to make them conductive. This is an essential prerequisite to the subsequent electroplating operations[2].

Dry film, which has virtually replaced wet film and screen printing for fine line circuit tracks, is now laminated simultaneously onto both surfaces of the board in a pattern which defines the copper tracks. This stage is referred to as 'primary imaging'[3]. Since dry resist techniques gained general acceptance in the 1970s, processing times have reduced from several hours to less than one hour, and modern machines are almost entirely automatic.

Copper is now electro-plated to a thickness of about 20 μm onto the portions of the pattern which will become the tracks, followed by 10 μm of 60/40 tin–lead plating. In-line plating systems with gantry-style transporters are available to handle a wide range of board sizes. The raising and lowering of PCBs to and from processing tanks, and the control of timing and other process parameters are under computer control.

After stripping the dry film away, the exposed copper which has not been plated with tin–lead is removed by chemical etching (e.g. using ferric chloride or cupric chloride). Again this process can be in-line with those detailed above. The tin–lead is next made to reflow, a process necessary to achieve good solderability. Various techniques are available, such as passing the boards through an infra-red stoving chamber.

The solder resist or 'secondary imaging' stage covers all regions except where soldering is required. Again the mask patterns for this process can come directly from the CAD system. Silk screening is sometimes used, but increasingly dry film techniques are proving more popular, using a process very similar to that described for primary imaging. Here the designer of the board can help improve its producibility considerably. Large copper areas, such as ground planes which are very often essential in radio frequency circuits (e.g. Fig. 4.2) exhibit a tendency to poor adhesion with the solder resist film, which may bubble upon soldering. In most cases the electrical performance will not in any way be degraded if a copper 'grid' pattern (Fig. 4.3) is used instead. The surface irregularities help improve adhesion and, as a side benefit, provide for more even plating thickness of holes in copper-intensive and -sparse areas of the PCB.

Finally, before profiling, the component identifications and any batch code are silk-screen printed onto the board using inks which will not be removed by the subsequent soldering and cleaning processes.

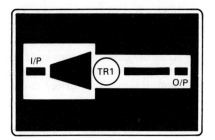

Fig. 4.2

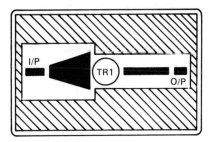

Fig. 4.3

4.4 ASSEMBLY OF PCBs AND HYBRIDS

Historically, the manual assembly of PCBs has been a costly and error-prone process. Not only can components be fitted in the wrong locations or wrongly orientated, but also physical damage may be done to their terminations during handling. This is particularly a problem where multi-lead components such as dual-in-line integrated circuit packages have to be inserted through holes in a PCB.

It will be apparent that, having defined the layout of the PCB, all the data on component selection, location and orientation is contained within this design data. All the information necessary for computer aided assembly can thus be derived from the design data by post-processing, and many CAD systems contain provision for this.

Digital circuits are particularly amenable to automatic assembly, since they generally consist of integrated circuit packs with perhaps a few discrete components such as pull-up resistors and decoupling capacitors. If possible when designing such circuits, it can be helpful to use thick film resistor assemblies mounted in packages similar to the integrated circuits. These are readily available in single-in-line or dual-in-line form, and so they can be handled by the chip inserting equipment. Decoupling capacitors generally do have to be single components, since they only act as an effective energy store if mounted very close to the particular chip to be decoupled.

Analogue and mixed circuitry can pose more of a problem, because large value capacitors and coils, for example, may only be available in radial lead form. It helps reduce production costs if component standardization is introduced at the design stage, and the establishment of a CAD component library acts as a discipline to discourage the selection of non-standard components except where really necessary. This is particularly important where leaded components are to be fitted through holes in a PCB, since a new tool may be required for the insertion machine. Surface mounted components generally pose less of a problem in this respect, since a range of component outlines can be accepted by a single pick-up head[4]. At present, however, not all component types are available as surface mounted devices (SMDs).

Ultimately it may be that on some PCBs there are one or two large components for which it is quite uneconomic to use computer aided assembly techniques, particularly if small batches of several types of boards are the normal production pattern. The degree of automation which can be justified depends upon such factors, as well as considerations of quality and response time of the manufacturing system to changes of component types. Therefore, in the paragraphs below, various levels of computer assistance for both inserted and surface mounting assemblies are considered.

4.4.1 Component insertion

Without computer assistance, an experienced assembly operator can insert typically 200–400 components per hour depending upon component types and degree of familiarity with the board layout. Multi-lead components, such as integrated circuits, can be rather difficult to insert manually, particularly if the leads have not been adequately protected and are misaligned prior to insertion.

Computer guided insertion

Simple systems which present the components in sequence to the operator and indicate the appropriate locations on the board can improve throughput, particularly where board types are changed so frequently that the operator does not gain familiarity with the layout. Typically, components move on a conveyor of small bins such that access to the appropriate bin appears via an aperture in the operator's workbench. At the same time the component location and orientation are indicated by light beams projected down from above the board. Alternatively a filmstrip or similar system can be used to provide this and even more detailed instructions if required.

The information for programming this type of assembly aid is contained in the CAD system, and appropriate post-processing software is generally available for the more popular types of machines. Compared with manual assembly, insertion rates are increased by typically 30%, but a significant saving may also be obtained as a result of reduced rework, since the number of assembly mistakes is reduced.

Semi-automatic insertion

Increased output is obtained using semi-automatic component insertion machines. Here the difficult task of fitting several leads through holes simultaneously is handled entirely by the machine. The operator is required to position the board beneath the insertion head and once it is correctly located the actual insertion is initiated by a press switch, which may be foot operated.

This type of machine is often selected for the difficult job of chip insertion. The sequence for all types of boards is made the same (Fig. 4.4) so that the operator, who needs to know nothing about the identification of the chips being inserted, simply moves through all chip locations in the prescribed sequence. Most of these machines have the facility to skip a location, which can be useful if component shortages prevent completion of the whole board.

A semi-automatic chip inserter can typically fit 1000 chips per hour, including due allowance for loading new components onto the machine and for occasionally clearing jams caused by damaged components. To minimize the chance of assembly damage the leads of each chip are held at the correct spacing by the insertion head. Underneath the board the projecting leads are crimped and cropped to length automatically, so upon

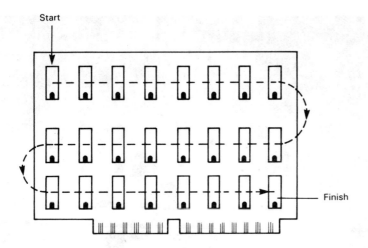

Fig. 4.4 Chip sequence for semi-automatic insertion

completion of assembly the board is ready for soldering and the components will not fall out of their holes if the board is inverted.

Automatic insertion

Fully automatic insertion machines need only to be fed with bare boards and components. The positioning of the board, the selection and insertion of the components is entirely automatic. 'Hit rates' (numbers of insertions per hour) depend upon machine and component types, but maximum rates of 6000 per hour are possible. When planning such a facility, it is important to take into account the time necessary for loading and unloading the machine, for clearing jammed components — which is still necessary from time to time — and of course for programming the machine ready for new designs of PCB. Thus a machine with a maximum hit rate of 5000 components per hour might typically achieve an average throughput of 3000 per hour (ten times the output of a good operator working entirely manually). Of course, if the boards have been designed on a CAD system with a suitable post-processor, then the tapes to drive the inserter can be produced without programming via a digitizer.

Machines are available to handle radial and axial lead components as well as integrated circuit packages. Very fast axial component inserters need to be fed with the components in a predetermined sequence. Machines for creating the necessary bandoliers of components are a further substantial investment, but are justified where high throughput capacity is utilized. In some of these machines, the components pass through an on-line tester to remove defective parts prior to sequencing. Some axial lead component inserters work directly from multiple magazines of components, thus avoiding the need for a sequencer.

4.4.2 SMA and hybrid microcircuit assembly

Because assembly techniques for SMA and hybrid microcircuits have so much in common, they are grouped together here. It is appropriate to recall that hybrids generally use ceramic or enamelled metal substrates, that they carry screen printed components (resistors and conductors) and quite often employ bare semiconductor chips connected to the tracks via micro-fine bonding wires.

*Fig. 4.5 Surface mounting
devices (Courtesy Mullard Ltd)*

SMA technology using small outline components provides increased packing density. This is partly due to the smaller physical size of the components (Fig. 4.5), many of which are used in unencapsulated form, but partly also because tracks can now be run beneath the components in regions previously taken up by lead holes. Components can if required be mounted on both sides of a PCB. However, the designer of an SMA layout needs to take even more heed of the actual manufacturing techniques which will be employed. For example, the design of the component pads should reflect the method of soldering to be used. This theme is further addressed in Chapter 5 when the subject of 'design for manufacturability' is discussed.

Components such as chip capacitors, small outline semiconductor packages, etc., are mounted using equipment identical with surface-mounting 'pick and place' machines. These machines hold the components by means of a vacuum tip which is manipulated either manually or automatically.

Semi-automatic surface placement

For mixed technology (surface and inserted components) and low production volumes, semi-automatic placement or 'onsertion' may be appropriate. Mixed technology may be forced on an OEM when suitable surface-mounting devices (SMDs) are not available or provide insufficient power handling capability. Mixed technology assembly in fact provides very high packing densities whilst still involving soldering on only one side of the board — an obvious simplification over surface mounting on both sides of a substrate, when some method of retaining the components on the first side is necessary while the second side is being soldered.

Suitable semi-automatic placement machines are available, and quite simple machines can mount 1000 SMDs per hour once the operator has become familiar with the machine. To secure the components prior to soldering, epoxy adhesive dots are placed in the required locations, either individually by means of an air pressure syringe, or *en masse* by means of a 'bed of nails' which is dipped in the adhesive and then lowered into contact with the board. Alternatively, the components may be held in place by means of sticky solder paste which can be dispensed by syringe or by silk-screening.

Typically, the method of operation of these machines is in mode reversal compared with semi-automatic inserters: the machine manipulates the board until the correct component location is indicated in a window on the computer screen. The correct component is presented in its bin in front of the operator who picks up the SMD using a vacuum pen and places it accurately into position, monitoring the process via an image on the screen. The image, from a video camera, is presented several times full size.

Programming such machines generally consists of a 'machine-learning' process in which the operator moves the board via the machine's $X-Y$ table, until a component location is exactly centred on cross wires on the monitor. Then data are entered via the computer keyboard detailing bin number, polarity and identification code as necessary. It will be noted, once again, that this data should be available from the CAD system with appropriate post-processing.

Apart from the component bins, some machines have facilities for feeding components from tapes or from stick magazines — the latter being used for transistors and integrated circuit packages. Anyone having a cold and being likely to sneeze is advised not to gaze closely into the bin carousel, since many bare chip components carry no identification codes and once mixed up they can pose a bit of a problem.

Automatic surface placement

Surface placement is essentially a much simpler operation than component insertion, and despite the smaller component sizes and high positional accuracy, automatic 'pick and place' machines achieve very high throughput rates with a single placement vacuum tip. Having essentially avoided the 'footprint' problem, it would appear feasible to use several placement heads simultaneously, and some of the more advanced machines do just that. Typically the placement rate for a single station is between 5000 and 15 000 components per hour.

4.5 SOLDERING PROCESSES[5]

Soldering of PCBs has been predominantly an automatic process for many years, and automatic soldering machines have provided improved consistency of quality compared with manual soldering provided that the process monitoring has been effective. The most commonly used soldering processes are briefly discussed below:

4.5.1 Wave soldering

The principle of wave soldering is illustrated in Fig. 4.6. Wave soldering can be used for certain types of SMDs, but the board layout must be such as to avoid 'shadowing': the solder wave, upon impacting one component, is reflected and a contact in the shadow region (Fig. 4.7) may not be soldered at all. This problem can also occur if board layout orientation and soldering direction are incompatible (Fig. 4.8) such that the wave divides around the component leaving a solder gap at the trailing edge. The back contact may not be connected at all. A more effective layout is thus to mount components

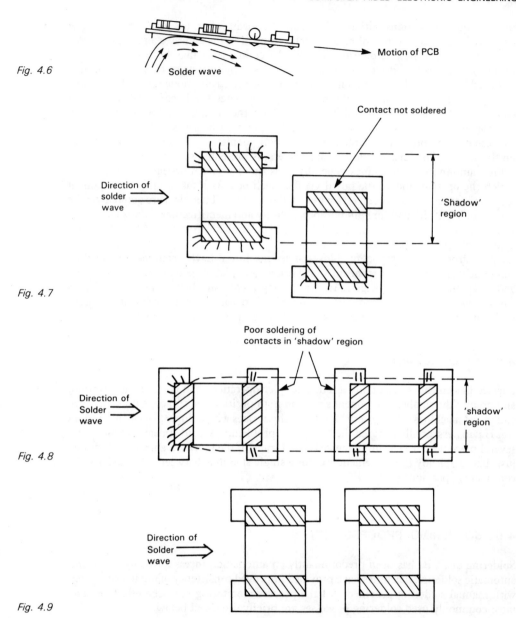

Fig. 4.6

Fig. 4.7

Fig. 4.8

Fig. 4.9

'sideways on' (Fig. 4.9) to the soldering direction. These types of layout rules can be built in to the CAD system software for SMAs.

A single wave is not the most effective for soldering SMDs, and dual-wave and jet-wave techniques — where the solder is pumped up into the air from a nozzle and returns by gravity — have been devised to solve the problems of 'solder skips' due to the shadowing effect.

Not all SMDs are suitable for flow soldering, and other techniques have to be used:

4.5.2 Vapour-phase soldering

Solder paste or cream, a mixture of solder and flux, is screened or syringed onto each location at which a component is to be soldered and the SMDs are placed (sometimes glued) into position. The loaded boards are placed in a tank containing a liquid whose boiling point is above the melting point of the solder. The liquid is heated and gives off vapour which condenses on the boards, giving up its latent heat and so raising rapidly the temperature of the surface mounting assembly. The solder melts and flows to form joints between the pads on the substrates and the SMDs.

4.5.3 Infra-red reflow soldering

This process again begins with a loaded assembly with solder paste at each intended joint. A conveyer carries the boards beneath heaters which first melt the flux and then the solder to create joints. Temperature build-up is more gradual, but unfortunately the components, being poorer reflectors of heat than the metallic solder, inevitably reach even higher temperatures than the melting point of the solder, and for some types of components this technique can cause damage.

4.5.4 Hot-plate reflow soldering

For ceramic substrates, it is common to apply the heat via the substrate itself. The loaded assembly is placed on a metal plate (or a heated conveyer) held at a temperature higher than the melting point of the solder. Ceramic is quite a good conductor of heat and so the solder paste soon melts and flows to create joints with the SMDs. This method of soldering cannot readily be transferred to glass-fibre substrates because they are very poor thermal conductors. To get the solder to melt, the underside of the board would have to be raised to such a temperature as to cause scorching of the substrate material.

Although these soldering techniques are the most commonly used, there are many other methods which have advantages in particular applications[6]. It will be noted that SMA technology commonly requires some if not all of the soldered joints to be made by reflow techniques. One particular problem with densely packed boards is that components can skew from their correct orientation (Fig. 4.10), with attendant increased risk of short-circuiting. A CAD system programmed with data for SMDs will also normally dictate the size and shape of the lands onto which they will be soldered in order to minimize these types of problems.

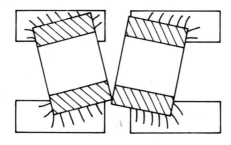

Fig. 4.10

4.6 INTERCONNECTION TECHNIQUES

The majority of interconnections within an equipment are likely to be made upon the PCBs carrying the electronic components. However, for interconnections between PCBs a 'mother-board' into which the 'child-boards' plug is often the adopted solution. Alternatively some form of wiring loom or cable form is required.

Bundles of wires — the traditional cable-form technology — are costly to make and to test. Ribbon cable, in which conductors are formed in a flat, parallel tape, are now more commonly adopted. The ends of the ribbon can be terminated by insulation displacement connectors (IDCs) so that cables can be manufactured automatically with no soldering involved[7].

One problem with ribbon cables is that conductors must run parallel to one another throughout and this places an additional layout constraint on PCBs. This restriction can be avoided by the selection of flexible circuits which are thin PCBs intended for use in situations where their shapes need to be altered. For example, the connecting cable between the daisywheel printing head of a Juki 6100 printer and its drive electronics is an 11-way flexible printed circuit. It is possible to use two or more layers of conductors so that cross-overs can be achieved if required. These types of interconnections can be designed on a CAD system, which can also produce the data for automatic testing of the cables. 'Flexi-rigid' PCB technology integrates conventional rigid PCBs with their flexible interconnecting cables. The components are mounted upon the rigid part of the board and the need for a separate connector or cable soldering stage is avoided.

The full benefits of surface mounting technology can hardly be achieved if the board has to be drilled and edge connectors screwed into place to allow connections within the equipment. Considerable progress is being made to develop surface-mounting connectors which can be fixed to the substrate using the same machines and processes as other surface-mounted components.

SUMMARY

Computer aided manufacturing and assembly systems are available for all of the main processes within a production unit. Computer control of individual equipment items, of process lines and of transportation of materials and products throughout the manufacturing plant contribute many benefits including increased productivity, more rapid response time to product change, and improved product quality.

At several stages in production of electronic equipment, data are needed to control the manufacturing and processing machines. In most cases this information can be obtained from the data created during design of the product. Most CAD systems offer facilities for post processing of the design data to create data for NC drilling and routing, photoplotting of artworks, and automatic insertion or placement of components.

At the present time there is no single accepted data format for CAD/CAM information interchange, and so appropriate data format translation must be incorporated into the 'engineering database' for the various production machines (Fig 4.11).

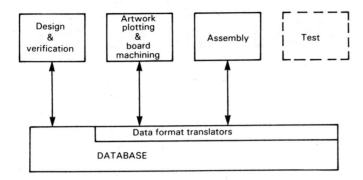

Fig. 4.11   Integrating CAD and CAM

REFERENCES

1. Cox, D. (1985) Printed circuits: mechanical cutting and drilling, and Routing or profile cutting. *Electronic Production,* **14** (4 and 5).

2. Scarlett, J.A. (1980) *Printed Circuit Boards for Microelectronics.* 2nd Edn. Electrochemical Publications.

3. Gorondy, E.J. (1985) A new in-line high productivity dry film solder mask process. *Electronic Production,* **14** (3), 20–34.

4. Down, W.H. (1981) Reliable soldering of chip components. *Electronic Packaging and Production.*

5. Comerford, M.F. and Swift, D.R. (1984) Soldering, cleaning and environmental testing of printed wiring assemblies that contain surface-mounted devices. Proc. P.C.C.W. III, Washington, D.C.

6. Comerford, M.F. (1979) Influence of circuit layout and lead configuration on the soldering of printed wiring boards. 3rd Annual Soldering Forum, San Diego, California.

7. Walton, J.P. (1985) Developments in interconnection equipment. *Electronic Production,* **14** (5), 9–15.

5 Computer aided manufacturing systems

5.1 OBJECTIVES

In this chapter, ways in which the CAM techniques outlined in Chapter 4 can be implemented are discussed, firstly by reference to typical equipment for manufacture and assembly processes and secondly by considering ways in which these facilities can be organized into production systems.

Some of the important considerations in the integration of CAD and CAM are introduced with particular emphasis on the requirements of *flexible manufacturing systems* (FMS).

In reading this chapter you should gain an insight into the essential features of:

— typical equipment for PCB and hybrid microcircuit production
— systems for materials storage, retrieval and transport within a factory,

together with an introduction to the subjects of:

— design for manufacturability
— integration of CAD and CAM
— factory automation systems.

5.2 IMPLEMENTATION OF CAM TECHNIQUES

In the paragraphs below, examples of current equipment used in computer aided manufacturing and assembly of PCBs and hybrid microcircuits are described. In the appendix to this chapter key performance parameters of a small selection from the very wide range of computer aided equipment for electronics manufacture are compared.

5.2.1 Machining operations

The machine illustrated in Fig. 5.1 is a combined drilling and routing machine from VERO Advanced Products Ltd. Under CNC control, and using data produced from the

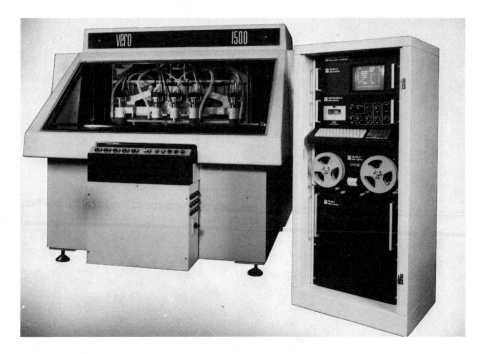

Fig. 5.1   VERO DR1500 CNC
drill/router

hole coordinates stored in the CAD system or from digitizing a master layout or PCB, this machine can drill 2000 holes per minute. It has special radial thrust bearings to make it suitable for routing, so that all aspects of the machining of a PCB can be completed on a single machine. A typical board can be fully machined in approximately one minute, including time allowed for loading and unloading boards from the machining table.

Automatic tool change facilities are provided, with up to 12 tools per spindle: tool change time is 15 seconds. Drill speed and feed rate are both under software control, with storage provided for a maximum of 30 sets of tool parameters (speed, feed rate, etc).

Tape preparation and editing facilities are provided as standard, with optional on-machine programming via an optics (or CCTV) system.

5.2.2   Assembly operations

In this section examples of equipments for conventional inserted component assembly and for surface-mounted assembly will be described.

DYNA/PERT dual-in-line inserter

Figure 5.2 illustrates the DYNA/PERT Model HPDI-F inserter for dual-in-line packages. The chips are mounted in plastic tubes which are fed via channels to the insertion head. This machine holds 7 magazines each with up to 15 channel positions, depending on lead centre dimension. The machine can accept PCBs up to 20″ × 19″ and can insert 0.300″, 0.400″, and 0.600″ centre components, straightening the leads in two dimensions prior to insertion.

Microprocessor control is built in for 'stand-alone' operation, but supervision can be

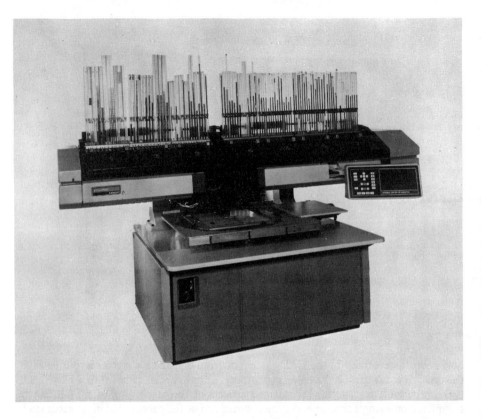

*Fig. 5.2 DYNA/PERT
automatic DIP inserter*

transferred to a central computer controlling many manufacturing operations involving other machines. Optional facilities include a component verifier which checks that the DIP component is the correct part and is orientated properly immediately before component insertion.

Up to 105 different types of dual-in-line packaged integrated circuit configurations may be selected at a cycle rate of 4250 components per hour, and PCBs with up to 300 insertion positions can be accommodated. Provision is made for expansion up to 1000 insertion positions.

A basic optical correction facility is provided as standard. This system, consisting of a light pipe and a photosensor, automatically corrects any positional error in location by reference to the actual position of the left front hole of an insertion location in a PCB.

The basic input/output device for the HPDI-F is a teletype terminal with tape punch/reader. Other I/O devices are available including a sophisticated data management supervisory system. Communication with a central computer is achieved via a 20 mA current loop or RS232C serial link. The standard software supplied with this machine includes the operation of the machine functions, insertion program editing function for minor editing of parts programs (more extensive editing can be carried out via an input/output terminal), machine diagnostics and management information.

CHIPLACER semi-automatic SMD placement system

The CHIPLACER (Fig. 5.3) is a semi-automatic placement machine for surface

Fig. 5.3 CHIPLACER semi-automatic placement machine for SMDs

mounting of all types of SMDs onto PCBs. Components may be loaded from bins mounted on up to three carousels each of which contains thirty component bins. Alternatively the components can be drawn from magazines or other types of SMD dispensers.

This machine incorporates an $X-Y$ table which moves the board under computer control until the required component location is accessible to the operator via a small aperture in the work surface. Positioning accuracy and repeatability of the table are specified as 0.05 mm. A small screen monitor provides a six times magnified image of the board so that the SMD can be accurately orientated and placed using a vacuum pen. Facilities for off-line programming are available and data are stored in a 32K buffered RAM which has power back-up protection for 60 days. Data can be entered and stored via mini-cassette tapes.

5.2.3 IV Products vapour-phase soldering system

The VPR 5000 series from IV Products provides a modular approach to building up an integrated preparation/vapour-phase soldering plant for SMAs. The pre-heat, vapour phase soldering and post-treatment de-fluxing stages can be integrated in-line and serviced via an overhead transporter under computer control. The system is fully automatic and provides built-in process audit and fault reporting. Process parameters can be programmed from the keyboard or via a serial data link. Several batches of work can be handled simultaneously even though each batch may require different process settings.

The installation illustrated in Fig. 5.4 is part of a flexible manufacturing system for SMAs being established at the British Aerospace Dynamics Group at Filton, near Bristol. Ultimately the project, which has received support from the UK Department of Trade and Industry, will involve robotic automation of virtually all aspects of the production of SMAs based upon PCB substrates, ceramic tiles and metal-cored substrates. The complete facility is to be controlled from a central host computer.

AUTOMATIC VAPOUR PHASE/CLEANING SYSTEM

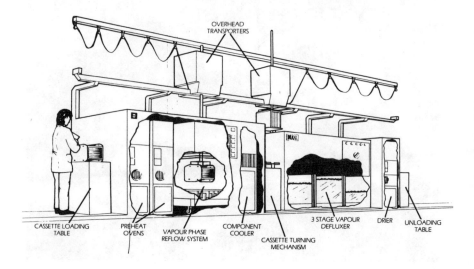

OVERHEAD
TRANSPORTERS

CASSETTE LOADING
TABLE

PREHEAT
OVENS

VAPOUR PHASE
REFLOW SYSTEM

COMPONENT
COOLER

CASSETTE TURNING
MECHANISM

3 STAGE VAPOUR
DEFLUXER

DRIER

UNLOADING
TABLE

*Fig. 5.4   Automatic vapour
phase/cleaning system*

5.3   DESIGN FOR MANUFACTURABILITY

As far as possible, the information in this book has been arranged in a chronological sequence, progressing through the life cycle of an electronic product. However, it must be stressed that some very important aspects of the work during both conceptual and physical design phases should focus upon considerations of design for manufacturability (and design for testability: see Chapter 7). Certain points have already been mentioned: for example selection of components types suitable for automatic insertion without the need for additional tooling. Some other factors are introduced below.

5.3.1   Design for automatic assembly

Several important factors should be considered when designing for automatic assembly. Few automatic inserters can cope with components at arbitrary angles, and for fastest assembly the components should be aligned. Similarly, the pitch between soldering pads for resistors and other axial lead components cannot be considered arbitrary since the assembly machine has to pre-form all such components prior to insertion. Thus the CAD component library must be related to the production standard adopted.

Chip-insertion machines have a finite 'footprint' which is bigger than the area of the chip itself. If another component is mounted within the footprint of the inserter, then it will prevent proper insertion of the chip and may itself suffer damage. Of course, some of these problems can be minimized by assembling components onto the PCB in a suitable sequence, but in general these constraints should — and can — be incorporated into the design rules of the CAD system. The checking software will then automatically flag up any potential violations of these rules before the design is finalized.

5.3.2   Design for flow soldering

One way in which the designer can help achieve successful flow soldering of high-

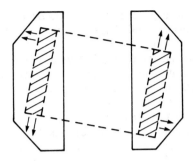

Fig. 5.5

density PCBs is by ensuring that tracks are produced only at minimum width and minimum spacing in one longitudinal direction on the board. Soldering should then be carried out in a direction parallel to these fine-line tracks in order to aid draining of excess solder from exposed tracks and hence minimize the risk of 'bridging' short-circuits. This constraint can be added to the design rules of a CAD system so that checking during design is carried out automatically.

The particular problem mentioned in Chapter 4 with regard to components skewing from their correct orientation (Fig. 4.10) can be alleviated by using suitable pad geometry. The surface tension of the molten solder can be used to pull components into line and to hold them there while the solder is cooling. Pads as in Fig. 5.5 can be used to help achieve this aim. Any tendency for the component to skew results in an increased surface-tension pull from the opposite direction, tending to correct the displacement.

5.3.3 Design for automatic handling

Care must be taken when designing PCBs for robot handling. Ideally, the robot gripper should be able to grasp an area at the edge of the board which is free of components. Similarly space left for a bar code, which can be screened onto the PCB, will allow automatic identification of board types. Thus the risk of trying to assemble components onto the wrong type of PCB — with the high risk of damage to the components and the board, if not to the assembly machine itself — is virtually eliminated.

Bar codes can be useful at various stages in the manufacturing cycle: a system of robot handling in an automatic test environment is described in Chapter 7.

5.4 INTEGRATION OF CAD AND CAM

The heart of an integrated CAD/CAM system is the database which stores the component information. This information consists of geometric data, performance data, and process related information.

5.4.1 CAD and CAM data

The nature of the data depends somewhat on the particular component and upon its position in the design hierarchy, as the following examples illustrate:

Example 1. *A binary counter primitive for standard cell integrated circuits*

Geometric data: Dimensional and shape data for each mask layer to be used in fabrication of the device.

Performance data: Functional and timing behaviour of the counter.

Process data: Statistical data on process yields, reliability for various manufacturing technologies, etc.

Example 2. *A thick-film DIL resistor array*

Geometric data: Package outline, pin coordinates, hole and pad diameters.

Performance data: Resistance values, pin connections, tolerances and power dissipation ratings.

Process data: Suppliers, costs, reliability data, etc.

These data are used selectively by various programs within the design and manufacturing environments: for example in simulation, physical layout design, mask-making and by machinery in manufacture and testing of sub-assemblies.

Many of the library data — such as component outlines — are essentially static in nature, rarely changing from when it is entered until it is replaced by data relating to a new component or process. However some data — cost and reliability are examples — need regular updating. Similarly, when a non-repairable component is found to have been damaged during an assembly operation, coincident with the generation of a module repair instruction, action must be taken to decrement the stock records by a further unit as this will be used in repair and must be reflected, in due course, in the re-order quantity. Thus if the parts purchasing system is also linked to the CAD/CAM database, this sort of data must be dynamic in nature, and yet it must be capable of being accessed by several programs.

Other data are created and used by the CAD/CAM tools in real-time: process monitoring and control involves such a system. Local memory and processing capability within the individual CAD/CAM tools is generally necessary for storing and manipulating these data. For example, some component placement machines incorporate a digital servo for automatic position correction. The optically derived control data are not used by any other sub-system within the manufacturing unit, although it may be necessary to process the data in order to extract management reports as input to a 'quality control' system.

5.4.2 Types of databases

Database design, data structures and database management systems (DBMS) are major subjects of study in their own rights[1], but it is appropriate at this stage to briefly discuss the dynamic nature of CAD/CAM data and the type of database most commonly selected for its manipulation.

Hierarchical tree and network structures

Figure 5.6 illustrates a tree data structure for electronic components. In this structure all decisions on routes through the tree have been reduced to binary forms. Other tree structures can be produced where the nodes have a different number of branches — or even a variable number[2].

Where low-level items are members of several branches the data structure becomes more complex. An analogue-to-digital converter (ADC) hybrid microcircuit may be

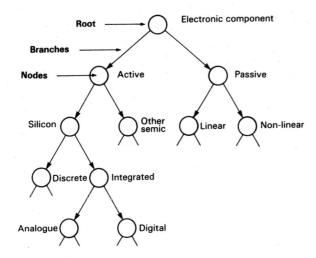

Fig. 5.6   Tree structure

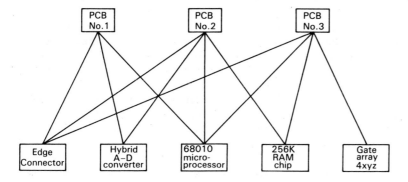

Fig. 5.7   Network structure

used as a component of two or more PCB sub-assemblies. This is reflected in the network data structure of Fig. 5.7. These sorts of structures are well suited to situations in which data are static or changed only infrequently. However, changes in one sub-assembly structure may necessitate alterations to the data structure or programs which make use of the shared data.

Relational Database (RDB)

Where regular data changes are envisaged — as is certainly the case in the early stages of a development project — a system in which data can be changed without altering the data structure or application programs is required. The RDB provides this flexibility.

Table 5.1 shows the structure of an RDB. The data files are arrays, each line entry being one 'record'. Attributes may be represented in one of several forms such as string expressions (e.g. suppliers' names and addresses), integers (e.g. unit quantities), boolean (e.g. preferred or non-preferred parts), decimal (e.g. hole coordinates in a PCB), etc.

There is no special sequence of records in a file, and new records can be added, old ones deleted or modified without affecting other data entries. New files can readily be created by selective manipulation of existing files. For example if a new electronic module is to be produced by modification of an existing design, it is a simple matter to

Table 5.1   Relational database structure

Attribute: Representation:	Description String	Quantity Integer	Preferred Boolean	Supplier String	£ cost Decimal
	Resistor ABC123	1	Y	Mullard	0.004
	Resistor XYZ456	2	N	Mullard	0.275
	Transistor 2Nxx	1	Y	Motorola	1.375
	.	.	.	.	.
	.	.	.	.	.
	.	.	.	.	.

Note: In this RDB file all records must be of the same type

program the DBMs to select a sub-set of an existing file as the basis for creating the new one.

5.4.3   Failure recovery

An important consideration when designing an integrated CAD/CAM system is its performance in the event of equipment failure. Should an individual process station fail, then clearly it will be impossible to maintain the output of that particular process — at least unless other similar machines are incorporated elsewhere in the facility. The effect on other stations of such a failure must be considered and, since most electronic assemblies are relatively small, it may be practicable to incorporate additional local storage facilities to cover such an eventuality.

The central computer itself could suffer failure, of course, in which case system operation would be very much dependent upon how much distributed processing power the system contained. Two such examples are discussed below.

Example 1.   *Workstation for logic chip design*
The workstation contains sufficient processing capability to handle schematics capture and simulation, using the central computer in batch mode only for the computation-intensive tasks of placement and autorouting. Thus the interactive design activities are unlikely to be delayed significantly unless central computer down-time is extensive.

If the system consisted of simple terminals connected to the central computer the designer would, of course, lose all service until the mainframe was brought back on-line.

Example 2.   *A DNC machine for PCB drilling and routing*
The CNC machine, having received its parts program from the central computer, can continue to produce PCBs even if communication with the mainframe is lost. It would not, however, be possible to change PCB types unless some back-up form of data transfer existed.

The ability of a CAD/CAM system to survive and rapidly recover from equipment faults is thus critically dependent upon the way in which processing power is distributed throughout the system[3], and upon the communication and process control software[4].

5.5   FACTORY AUTOMATION

As complexity increases and with it the need for greater precision in assembly, the computer aided manufacturing and assembly techniques discussed above may have to be adopted even for development prototypes and for 'hand-crafted' special purpose products built in very small quantities. Here, however, attention will be directed towards two levels of production which encompass the bulk of the output of the electronics manufacturing industry: mass production and batch production.

5.5.1 'Detroit' automation

Because of its origins in the motor industry, automation of high-volume production has been dubbed 'Detroit' automation[5]. Fig. 5.8 illustrates in block diagram form an in-line flow system suitable for manufacture of a single product type where there is little or no variation between processes applied to individual units of output.

If the system is to flow smoothly and machines are to be utilized efficiently, the processing rate at each station must be closely balanced. In manufacturing and assembly this can be achieved by putting several stations in parallel when a slow process is involved. Alternatively, a process such as PCB assembly can be split into a number of sequential operations with machines in series carrying out part of the task.

Not all processes can be treated in this way, however, and it may be necessary to collect items into batches before processing them. For example PCBs are usually stacked and pinned four high onto drilling tables, so that twenty boards are drilled simultaneously on a five-spindle CNC machine. Local storage is therefore necessary, with the boards being fed onward either as batches or in sequence depending upon the nature of the next stage of manufacture.

Planning of the assembly stage is somewhat simplified by the deterministic nature of the task: this is not so with testing, however. The time required to test a PCB can be predicted quite accurately, but if faults are detected, then a further amount of time must be allowed for diagnosis and creation of repair instructions. The rate at which good boards are passed on to the next stage of assembly is therefore subject to fluctuations which necessitate local storage of parts or non-synchronous transfer. Inevitably some machines are kept waiting for work.

Logically, then, one aims to fully utilize the expensive items of capital equipment, such as automatic assembly machines and functional testers. These machines determine the nominal rate of production, and this is achieved by providing sufficient low-cost stations to cope with most of the peak demands made upon them. Some of the lower-cost stations are inevitably under-utilized.

Flow-line production is employed in the manufacture of television receivers and similar high-volume products where hundreds of thousands of identical items are manufactured per year.

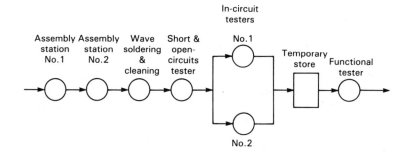

Fig. 5.8 'Detroit' automation line

5.5.2 Flexible manufacturing systems (FMS)

A flexible manufacturing system carries out operations under the control of a dynamic scheduling system. The actual processes and in most cases the equipment used are identical to those already discussed above. The difference between an FMS and a flow or transfer production system is that in an FMS the scheduling of activities is carried out on-line, with work being transferred between processing 'cells' of intelligent machines capable of performing a range of manufacturing or assembly tasks determined by the control software.

Thus, for example, if the 'first-time pass' rate in testing one module of a piece of equipment falls substantially below normal, the resulting bottleneck in fault diagnosis can be prevented by re-allocating another test station to fault diagnosis work until the short-fall has been overcome. Similarly if a cell breaks down, processes within other cells are re-allocated and transfer routes are changed automatically so that machine utilization and throughput are optimized.

Implicit in the above description of an FMS is the need to be able to program the functions of individual cells and to control the transfer of materials and the communication of data between cells.

Materials storage, retrieval and transfer systems

Since the FMS must be capable of adapting to changes of product type, the materials fed into the manufacturing unit must also be controlled from the FMS database. Automatic storage and retrieval systems are available and are used extensively in mechanical-engineering production[6]. These systems are, of course, equally suitable for handling mouldings, racks, castings and other large components of electronic products. Small electronic components are available in packaging suitable for automatic handling. Examples include bandoliers of axial lead components and blister tapes of SMDs. One person can normally keep a number of automatic insertion machines loaded with components, but the decision as to which types of components to load is made by the FMS controller software.

Transfer of components and partly finished goods within an FMS can be achieved by automated guided vehicles. These are driverless tractors which receive their directions from the central computer either via buried underground cables or by radio. The vehicles keep to prescribed tracks but their routes can be altered in response to the changing situations. Since carts are generally identical one with another, it is possible for any idle cart to take over a route if the vehicle assigned to that route needs to return to a battery-charging station. Vehicle provision is another statistical calculation, but the FMS controller assigns task priorities in the event of a temporary shortage.

The future of FMS

The first fully integrated electronics design and manufacturing systems of this type have only recently been developed[7,8] and there remain many obstacles to their general implementation. It is the author's opinion that such systems will be established by larger companies at an increasing rate during the rest of this decade, and that smaller companies will need to 'buy time' on these systems if they are to achieve market competitiveness in the future. FMS brokerage may emerge in the early 1990s much as silicon brokerage is developing at the time of writing this book.

SUMMARY

Automatic systems for manufacture and assembly of electronic products provide increased quality through accuracy and repeatability as well as dramatic increases in productivity. Best results are only obtained from these systems if their use is considered right at the outset of design.

A database system is the heart of an integrated CAD/CAM system. For most types of products, flexible manufacturing systems provide adaptive production environments where high productivity is complemented by the ability to respond to changes in demand or to changes of product specification. Such systems are inherently more robust than conventional and 'Detroit' automation systems, since they are capable of reconfiguring in the event of partial system failure.

REFERENCES

1. Martin, J. (1977) *Computer Data-base Organization*. Prentice-Hall, Englewood Cliffs, New Jersey.
2. Ranky, P.G. (1983) *The Design and Operation of FMS*. IFS (Publications) Ltd, Bedford
3. Duffie, N.A. (1982) An approach to the design of distributed machinery control systems. *IEEE Trans. Industrial Appns*. **IA-18**(4), 435–41.
4. Groover, M.P. (1980) *Automation, Production Systems, and Computer-aided Engineering*. Prentice-Hall, Englewood cliffs, New Jersey.
5. Weidlich, S. and Prutz, G. (1982) Selection criteria for the use of digital distributed control systems. *Process Automation*, **2**, 66–74.
6. Hartwig, G.C. (1982) Computer aided warehousing. *Manufacturing Engineering*, **88**(2), 79–80.
7. Allderdice, H.B. and King, R.I. (1985) The design of a computer integrated electronics manufacturing system. *Computer-Aided Engineering Journal*, 57–65 (April)
8. Cronin, D.E. (1984) Computer aids and their integration. *Computer-Aided Engineering Journal*, 61–5 (February)

APPENDIX A5.1

Some features of a selection of current CAM equipment are compared in the following tables. It should be noted that these represent only a very small sample of the range of existing CAM equipment, and are provided as an indication of the capability of current technology.

Table A5.1 Typical component insertion equipment

Supplier	Product	Application/features
Amistar	AI-6448	Fully automatic component inserter for axial lead components. Maximum rate 9000 components/hr. Optional component verifier. Self-programming with storage of programs on floppy disk.
CONTAX	CS-301A	Automatic DIP inserter. Typical average rate 2000 components/hr. Integral programming facilities.
Universal	6295A	High-speed automatic inserter for axial lead components. Dual insertion heads give cycle rate of 30 000 components/hr.
Universal	2596A	Component sequencer for axial lead parts. Input modules can be added to accept up to 160 component types. Rates up to 25 000 components/hr.

Table A5.2 Typical SMA/hybrid assembly equipment

Supplier	Product	Application/features
Amistar	SM-1000	Fully automatic placement of SMDs. Maximum rate 14 400 components/hr. Optional polarity and value check prior to orientation. Automatic board loading/unloading facility available.
Mullard	MCM-III	Fully automatic placement of SMDs. Hardware-controlled multi-station system. Maximum rate for a four-station system 184 000 components/hr. Up to 12 combined stations maximum.
Universal	LOGPLACE	Semi-automatic SMD placement using light-beam guidance. Can be linked to a central computer. Vacuum-pen placement. $5\frac{1}{4}$ inch disk storage for programs.
Hughes	HMC-2460	Fully automatic semiconductor chip bonding machine for hybrids or SMAs. Typical time to bond a 50 IC circuit is < 5 minutes cf. 1 hr by manual means. Optional pattern recognition system.

6 Computer aided testing techniques

6.1 OBJECTIVES

This chapter focuses upon the philosophy and techniques for testing electronic components, assemblies and complete items of equipment. The philosophy of testing is discussed and various test strategies are considered. Finally, the problems of diagnosing and rectifying faults are analysed and some computer aided repair techniques are outlined. In Chapter 7, ways in which these techniques are implemented in particular automatic test systems, and their application in a production environment are discussed.

In reading this chapter you should gain an insight into:

— testing philosophies and their relationships with initial defect rates and costs of failure
— techniques for component testing
— techniques for testing electronic assemblies and systems
— techniques for computer aided fault diagnosis and repair.

6.2 TEST PHILOSOPHY

Every test carried out upon a product adds to its cost, delays its availability to the customer and, if no faults are found, adds nothing to its value to the end user. Testing is at best a necessary evil brought about by the possibility of error. At worst — when it fails to reveal faults which are present — it *can* be a complete waste of time and money.

When one considers the testing of electronic components, the possible outcomes are illustrated in Fig. 6.1. If the components are non-repairable (e.g. semiconductor devices) a simplified decision tree applies, and by inserting probabilities and costs at each possible outcome an optimum amount of testing can be ascertained. This may appear somewhat over-simplified, for while it is not too difficult to quantify the cost of increased testing in terms of capital expenditure, running costs and work in progress, it is more difficult to estimate the likely cost of lost future business through delivering faulty product to the customer. When reworking faulty assemblies is a possibility, the decision tree can be used to evaluate the economics of so doing.

In practice, OEMs are very reluctant to throw away complex electronic assemblies

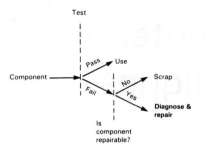

Fig. 6.1   Component test decision tree

and hence a repair strategy is adopted in many instances where it cannot be justified from financial considerations alone. Other factors of importance are the time delay and difficulty of replacing a small percentage of components if reject rate exceeds forecast.

Two types of test philosophy already emerge[1], viz:

— GO/NO-GO testing, for non-repairable items.
— Diagnostic testing, for repairable items.

Even this is rather a simplification, because a component manufacturer will certainly wish to understand the mechanisms and reasons for failures if the occurrence is sufficiently frequent, in order that corrective action can be taken: this type of activity can be undertaken outside the production test function.

For an electronic assembly there is a range, or 'spectrum', of types of faults. Some are due to bad components, others to errors in assembly or to faults introduced during manufacturing processes (e.g. solder bridges or open circuits.) Testing itself carries some risks, particularly when fault diagnosis necessitates probing to monitor voltages within an assembly. The form of a fault spectrum is illustrated in Fig. 6.2, although the percentages in each category are very much dependent upon the type of equipment considered. The nature of the circuit, the processes used for manufactured, and — not surprisingly — the purchased parts quality level all have an influence upon the ratios for each category of fault. In the above discussion we have *assumed* that the design has been thoroughly 'toleranced' (performance verified over the range of anticipated component tolerances), and that if all components are within specification the assembly will meet its overall performance requirement.

Fig. 6.2   Typical PCB fault spectrum

6.3 COMPONENT TESTING

The parts which make up an electronic assembly or equipment are likely to comprise:

— electronic components (semiconductors, capacitors, front panel controls and connectors, etc.)
— bare PCBs and back-planes or mother-boards
— interconnecting cables, leads, etc.
— mechanical housings, brackets, etc.

Apart from dimensional inspection, testing of mechanical items is usually restricted to destructive tests on a very small sample; so here attention will be concentrated on the other three categories.

6.3.1 Testing electronic components

It might be expected that the manufacturer of components would carry out tests on each delivered item. In some instances this is the case, although it is rare for every individual component to be tested to all aspects of its specification since this involves considerable expense. Normally, every integrated circuit is probe tested at the wafer stage, but a lot more happens to a chip before it is ready for shipping. Components for the most critical applications, such as satellite electronics, submarine cable repeaters, etc., where the cost implications of failure are horrific, are put through rigorous screening tests to verify not only that they function to specification, but also that they can withstand the most severe environmental stresses without failure or unacceptable performance degradation.

For the majority of components, the pre-delivery testing is likely to be limited to all — or more commonly just a small percentage — of the parts being subjected to basic performance tests at normal room temperature, with a sample from every production batch being assessed more thoroughly as a quality control measure [2]. (BS9000 in the UK and MIL Specs in the USA involve a system of sample testing.) Much depends upon the specification against which the components are purchased, and it is always possible to pay a higher price to obtain a more thoroughly screened source of supply.

Semiconductors provide a very good example of the way in which some types of components have improved in quality in recent years. Each year, for example, National Semiconductor set themselves improvement targets in terms of numbers of defective devices escaping detection during test. For power semiconductors — a difficult manufacturing technology — failures were well below their target of 400 parts per million in 1984, and for the succeeding two years targets of 300 ppm and 200 ppm were established.

6.3.2 Justification for component testing

In the semiconductor industry there is a saying that 'any devices which *can* be made ultimately cost $5 — except those which cost less!' The relatively low individual cost of most types of electronic components might appear justification for accepting the occasional component defect: on average typically between 1% and 5% of commercial grade components are faulty on delivery[2]. However, it is necessary to consider the cost of finding and replacing a faulty component, and this depends on how late in the product life cycle the fault reveals itself, as Fig. 6.3 shows. It is difficult to quantify the financial implications of gaining a reputation as a *supplier of poor quality goods*.

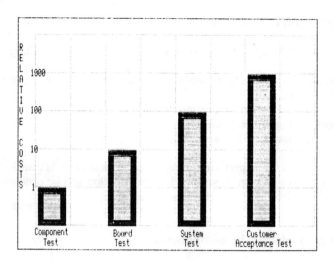

Fig. 6.3   The 'rule of tens' law
           for costs of failures

At each stage (component, assembly, equipment) the test policy should be directed at minimizing overall cost. It might appear from Fig. 6.3 that if less than 10% of components are faulty then it is more economic to omit component testing and to correct all faults at the next level up; at board test. There is certainly a level of faults below which component testing cannot be economically justified. It is instructive to consider the probability, P, of building a fault-free assembly as a function of component count, n, and quality level; the latter expressed as a component fault occurrence, F. From simple probability theory,

$$P = (1 - F)^n$$

Fig. 6.4 is a plot of loci of constant P versus F and n. It will be noted, for example, that to obtain 75% first-time passes at board level with a component count of 100, the component defect rate must not exceed 0.3%. At a board complexity of 200 components the initial component defect rate would have to be *below 0.15%*. In practice, with realistic purchased component quality levels, this would of course mean that virtually every

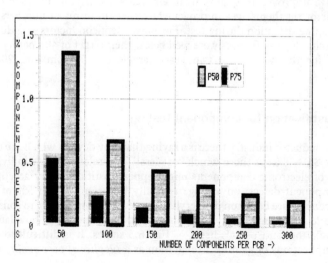

Fig. 6.4

board would fail. In fact doing the sums for this second example with 2% component defects reveals that 98.25% of boards would contain one or more faulty components. It would not be necessary to *test* the boards. . . It would be more economic to put them *all* straight into fault diagnosis and repair.

The options exist for the OEM to do:

— no component testing at all
— sample testing
— 100% testing of components.

Other options do of course also exist: for example by testing active devices only, but taking the above simplified situation and assuming that component input (i.e. purchased) quality is fixed at fault occurrence F, one can evaluate the relative merits of 100% testing, sample testing or zero testing as follows:

Let s = sampling fraction (typically 0.05)
  F = fault occurence factor (less than unity)
  C_1 = cost of testing a single component
  C_2 = cost of finding and fixing a fault at board test
  C_3 = cost of finding and fixing a fault at equipment test
  C_4 = cost of finding and fixing fault after delivery
  P_2 = probability of finding fault at board test
  P_3 = probability of finding fault at equipment test
  P_4 = probability of finding fault after delivery.

Note: It is assumed that $P_2 + P_3 + P_4 = 1.0$ since any faults existing but which are not found at all do not imply failure in terms of a financial penalty.

The average cost of finding and fixing at a later stage a fault which escapes detection at component level is:

$$C_L = \sum_{n=2}^{4} C_n P_n$$

The per component cost of zero testing at component level is thus

$$FC_L$$

The cost of sample testing is

$$sC_1 + (1 - s) FC_L$$

i.e. the sum of the costs of sample testing plus the cost of dealing with any faults which show up later. It is assumed that faulty batches of components are rejected and returned to the manufacturer.

The cost of 100% component testing is

$$C_1$$

Note: Fault occurrence factor, or average output quality level (AOQL), for supplier sample-tested devices can be ascertained from acceptance quality level (AQL) tables if batch size and inspection level are known. It should not be assumed that 0.65% AQL means 99.35% of components are likely to be faulty. AQL relates to the supplier's risk of rejection of a batch of components. Depending upon the batch size, the actual percentage of faulty components could be significantly larger or smaller than this figure [2].

Example

Putting some values into the above, assume:

$$s = 0.05; \quad F = 0.01$$
$$C_1 = £0.10; \quad C_2 = £1.00; \quad C_3 = £15.00; \quad C_4 = £300.00$$
$$P_2 = 0.82; \quad P_3 = 0.1; \quad P_4 = 0.08$$

and consider an OEM who uses 1 million components per year. The annual cost of each of the above strategies would be:

Zero testing: $£\{0.01 * (0.82 + 0.1 * 15 + 0.08 * 300)* 10^6\}$
                 Cost = £263 200.00

Sample testing: $£\{0.05*0.1 + 0.95 * 0.01 * 26.32\}* 10^6$
                   Cost = £255 040.00

100% testing: $£0.1 * 10^6$
                 Cost = £100 000

A substantial cost saving is forecast if all components are tested. Of course, actual figures of costs and probabilities for a real situation differ somewhat from the above, but at least for this example we can see that component input quality would have to be improved to well below 0.5% defects before component testing at 'goods inwards' became non-economic.

Many OEMs now adopt a policy of 100% goods inwards testing of all electronic components. The nature of the tests applied depends upon the type of component. Passive devices can generally be tested quite simply by value measurement, as for example would be appropriate for ceramic capacitors. It may also be necessary to apply full working voltage and to check leakage resistance in some cases.

6.3.3 Component reliability

The best way of finding a faulty component is to test it *as a component*. An effective automatic component tester would normally be able to detect well over 95% of faulty components. The success rate at higher levels (i.e. board test and equipment test) will be progressively lower. This is due in part to the fact that board and system tests do not stress each component to the limit of its performance specification, so that marginal or ailing components may escape detection during higher level testing. Some of these component 'faults' may never cause trouble to the end user of the equipment, while others may only show up in aspects of the operational role not evaluated at board or equipment level; or they may cause system failures in the field only after a period of time due to continued degradation.

6.3.4 Burn-in testing

The familiar 'bathtub' curve [3] of failure rate against time for electronic components (Fig. 6.5) shows that while component quality may be related to initial defect rates, reliability is a function of time. Some components are incorrectly manufactured and would never work, but many failures — particularly of semiconductor devices — occur after a period of operating time [4]. For some of these the few microseconds or milliseconds of warm-up time as power is first applied may be sufficient to cause catastrophic failure. For devices having inadequate thermal bonding to their headers the time constant may be long enough for the chips to survive automatic component testing, so they

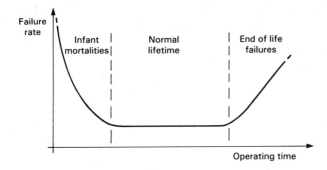

Fig. 6.5 'Bathtub' curve

may be passed through for assembly and subsequent board level testing. Other wear-out mechanisms, such as those due to electromigration, are also temperature and time related, but these types of failures may only show up in the field unless something can be done to accelerate the ageing effects [5]. It is found that semiconductor infant mortalities — once the initial defects have been screened out — involve lifetimes of thousands of hours (months of normal operation in the field) at comfortable room temperatures. If we operate the devices at elevated temperatures, the mean time between failures (MTBF) of these weaklings reduces to more manageable proportions. Typically it is found that lifetime halves for every 10°C increase in ambient temperature. Thus a device with a 10 000 hour lifetime at 20°C would fail on average after only

$$\frac{10\ 000}{2^{10}} = 10 \text{ hours at } 120°\text{C}$$

If we assume a constant percentage failure rate for these infant mortalities, it follows that failures will reduce exponentially with time, and operation for 24 hours at 120°C thus leaves only

$$100 * e^{-24/\text{MTBF}} = 9\% \text{ of the weak components}$$

undetected.

24 hours may not be an unreasonable amount of time for which to subject at least the most critical components to high-temperature operation, or 'burn-in'. This type of screening is carried out using oven-mounted panels pre-wired with sockets into which the components are fitted. Bias is applied and the components are generally monitored for catastrophic failure. If a device fails it can be manually removed from the panel or automatically disconnected from the source of power. On completion of burn-in the components are tested in the same way as if they were entering directly via 'goods inwards'.

6.4 TESTING ASSEMBLED BOARDS

Visual inspection of an assembled PCB may or may not be carried out depending upon the nature of subsequent testing. However, inspection by human operators varies greatly in effectiveness and while people are generally good at monitoring overall cosmetic appearance, they are more variable in their ability to notice manufacturing defects such as damaged tracks or incorrectly fitted components. The first computer pattern-recognition systems for this type of work are now being introduced, and their use will become more general in the near future. If the next stage of testing involves operating the board with normal working voltages applied, then thorough visual inspection will probably be justified, since one wrongly fitted component could cause damage to several others and involve a costly repair action.

The types of tests applied and the order in which they are done are termed 'testing mix and strategy' [6,7]. Before going into this subject it is appropriate to consider the two main testing techniques for loaded printed circuit boards.

6.4.1 In-circuit testers

An in-circuit tester (ICT) effectively acts as an automatic inspection machine, checking that each component has been fitted in the correct position and orientation, and that it functions as it should. The philosophy is based upon the premise that if everything is wired up correctly and if all the parts work individually, then the board as a whole should also function correctly.

In operation, the ICT consists of a 'bed of nails' (Fig. 6.6) matched to the layout of the particular PCB. Each nail is in fact a spring-loaded probe and is connected via an interface to driving and sensing circuitry. There may be typically 1500 of these board probes and the more advanced ICTs have facilities for programming the role of each pin, with perhaps a small number of pins reserved for special functions (e.g. very high-speed logic lines.) Once the interface jig containing the bed of nails has been made and fitted to the ICT, boards are loaded sequentially and held in place by a vacuum system which pulls the board down onto the probes. The ICT is generally used twice on each good board passing through test (faulty boards may of course be re-cycled through ICT after repair): first with no supplies connected to the unit under test, in order to check for short-circuits and gross errors in component values, and for a second time with supplies on the board, this time checking the operation of each component in turn.

When the ICT finds a fault it provides a diagnostic output which in most cases pin-points the location of the trouble to a particular component, or at least to one specific

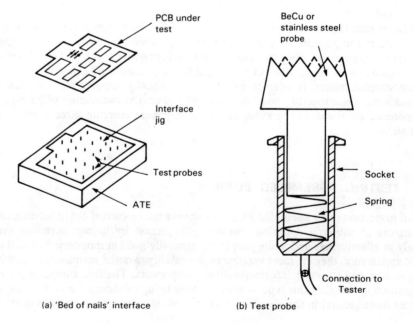

Fig. 6.6   In-circuit tester   (a) 'Bed of nails' interface   (b) Test probe

node on the board. There are cases where a short-circuit fault at one location may mask other faults nearby, but generally an ICT can be used for the diagnosis of multiple faults — it is not necessary to halt the test sequence and to fix the first fault before proceeding with the test run. The decision whether to repeat ICT after repair is an important aspect of test strategy. It may, for example, be appropriate to route all boards repaired after a single fault directly on to the next level of testing, while boards with multiple-faults are repaired and re-cycled through ICT.

Some types of faults are not generally detectable by in-circuit testing. These include failures due to timing errors and interactions between components. The time taken to ICT a board is generally very short (typically less than a minute for a good board; a faulty board takes longer only by the amount of time needed to print-out the diagnostic report) compared with the lifetime of 'infant mortality' weakling components, so very few of these are detected during ICT. Experimental evidence suggests that between 1% and 5% of semiconductor devices can exhibit 'infant mortality' characteristics, so the quality of products leaving the ICT may depend a great deal upon policy towards burn-in [8].

The interface jigs necessary for each different PCB type are an expensive aspect of the running costs of an ICT. The data for their manufacture can come from the CAD system, and some PCB design systems now contain post-processing packages via which NC machining of the ICT interface jig can be initiated. (Figure 2.9 illustrates the hole-location data in graphical form for a simple PCB. Hole-diameter information comes from the component library.) The alignment of the board on its jig is crucial, and can constitute the major time element in an ICT cycle. If good probe contacts are not achieved, then results are meaningless, so boards must be thoroughly cleaned before mounting onto the ICT. Solder flux with a low solid content has been developed to help minimize this type of problem.

The close pad spacing on SMAs places additional constraints on test interface design. Probes for mounting on 0.050 inch centres are more fragile than the probes traditionally used on ICT interfaces, and board positioning becomes even more critical [9,10].

The quality of an ICT system, while partly determined by the power of the hardware (its execution speed, ability to drive complex chips such as microprocessors and memories, etc.), is mainly determined by the software functions provided. The ICT software determines how easy it is to program the machine for a new board type: some ICTs have automatic test-generation facilities, making the writing of ATE programs relatively simple, particularly as the tests themselves, being on individual components, are generally fairly simple. The fault diagnosis and repair facility of an ICT is almost entirely determined by its software.

6.4.2 Functional testers

Functional testers (FNTs) subject assembled boards to a series of tests closely simulating the operational role when boards are mounted in the final equipment. Consequently FNTs give the highest assurance that a board, having passed its tests, will operate satisfactorily in the final equipment. These types of testers provide very high fault coverage but can only find one fault at a time so, if multiple-faults exist, several repair/re-test cycles via an FNT may be required before the board is fault-free.

The accuracy with which faults can be pin-pointed using an FNT is rarely as good as with an ICT, so that fault diagnosis generally involves some operator intervention. Upon detecting a failure the FNT provides a series of instructions to an operator who is directed to move a probe to various nodes of the circuit. In some computer aided repair (CAR) systems based upon functional testing, the board layout is displayed upon a colour monitor with the required probing points indicated. When the computer has located the faulty component it is highlighted in a contrasting colour on the display and a repair instruction label is printed.

To generate the complex stimuli and to measure the responses of the unit under test, programmable instruments are part of an FNT station. The operation of each instrument is controlled by a computer which organizes the test sequences. Connection to the unit under test is generally by means of the edge connectors of the board, so that access to internal nodes is available only via the 'guided probe' during fault diagnosis. For digital boards the necessary pattern generators are included in the ATE itself, but when analogue or mixed circuitry is involved several other types of signal sources or monitors may be needed.

Fortunately, the introduction in the 1970s of the IEEE 488 standard for bus-compatible test equipments and controllers has resulted in a wide range of test instrumentation to choose from [11]. This General Purpose Interface Bus (GPIB) standard, originally pioneered by Hewlett Packard as HP-IB, not only specifies mechanical dimensions and pin allocations for plugs and sockets, but also defines handshaking, control and data formats and lays down timing requirements. Thus with little difficulty one can configure an FNT using a range of programmable instruments from several manufacturers.

With low-frequency circuits (e.g. audio systems) it is possible to operate the functional tester at the speed of the board in its operational role, hence giving the unit under test a true 'performance test'. High-frequency digital systems cannot be tested at full speed on an ATE and dynamic testing of such boards is thus carried out at a 'keep-alive' speed with which the computer can cope. Some FNTs can test digital systems at up to 50MHz or more, but this is still a long way short of true operational speed for many digital systems. Thus a risk remains that some faults, such as those due to excessive time delays or slow switching characteristics, will not be found during FNT.

The interface between a board and an FNT is usually just an edge connector; a substantial cost saving compared with the ICT. For boards containing microprocessors an additional cable is used to connect the microprocessor chip socket via an 'emulator' to the FNT [12].

Programming on the other hand usually takes rather longer for an FNT (although FNTs with automatic test-generation facilities are now becoming available).

It should already be clear that there are advantages and disadvantages to both ICT and FNT. For very complex logic boards overall testing on an FNT is rarely effective in isolating faulty devices: it has become normal practice to write programs to test small functional areas of the board at a time. If these areas become so small that only a single component is being tested, one is effectively back to ICT. ATE systems combining ICT and FNT are now available and can offer advantages in certain applications: there is no longer the need to transfer boards between the ICT and FNT, so that loading and unloading times are greatly reduced and all boards can travel the same route through the test department whatever testing strategy is adopted for them.

There are, however, some disadvantages of the combined ICT/FNT approach: the equipment tends to be very expensive, and for many boards 50% of the facility may not be needed. The test interface is even more expensive than for ICT, since it must be designed to minimize cross-talk between the numerous ICT probe contacts. Even so, the speed of such combined testers during FNT is inevitably limited due to the large amounts of capacitance associated with the connections to each node.

6.4.3   Testing mix and test strategy

In planing a test facility, particularly when expensive ATE is involved, the decisions on which types of automatic testers should be used, and how exactly they should be utilized to best advantage, can have a substantial effect on the cost of production and the ability to meet output demand.

Testing mix

This is a term used to describe the test equipment available. It defines the different types of testing which can be carried out within a test facility, and is dependent therefore on the nature and quantity of capital investment. For example, a small facility might be limited to a functional tester at board level plus board substitution at equipment (or 'system') level, whereas a comprehensive test facility might have an available testing mix comprising component testers, burn-in chambers, continuity testers, ICT and FNT at board level and perhaps even a special-purpose automatic system test-set.

Test strategy

For a particular product, the test strategy defines the way in which the testing mix facilities are applied for that product. Thus the test strategy may differ for each board in a system in whether or not it calls for a particular type of testing and also in the sequence in which tests are carried out (the route through the testing mix.)

Some types of boards might be most economically tested using the strategy indicated in Fig. 6.7, for example, whereas the strategy of Fig 6.8 would be more appropriate for other types. In perhaps the majority of cases the strategy of Fig. 6.7 turns out to be economic, particularly if a high proportion of boards are expected to fail first time round. The reduced fault diagnosis time of the ICT means that throughput can be maintained despite the high initial defect rate.

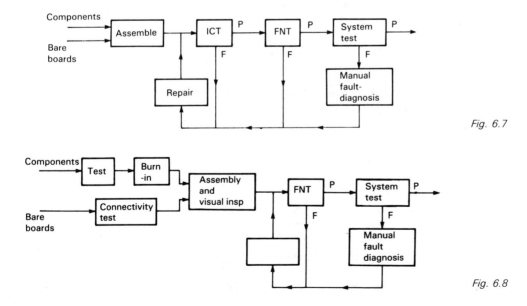

Fig. 6.7

Fig. 6.8

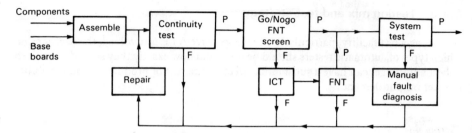

Fig. 6.9

When high-quality components are assured — by 100% testing at 'goods inwards' — and the majority of infant mortalities have been induced by burn-in testing, it may be that the expected number of faults per board is well below unity, so that the majority of boards would pass the functional test first time. In this case the ICT should not normally reside in the forward path of the testing flow chart. Indeed it may not be justified at all or, as in Fig. 6.7, it may only be used in the fault diagnosis and repair loop where its faster fault diagnosis capability compared with FNT will be advantageous.

Clearly, one could postulate several testing strategies, from simply building boards and plugging them into a system (just as is often done when testing analogue integrated circuits), to very complex test schemes with ICT and FNT used as in Fig. 6.9.

6.4.4   Selecting a test strategy

The task of selecting an appropriate test strategy can be formidable, particularly when several products are competing for time on some of the test facilities. Meaningful financial appraisal of various strategies must include such important features as:

— present loading on each tester
— production rates for each board type
— rate of 'field returns' anticipated
— working plan (shifts) in test department
— cost of each tester
— annual maintenance costs (e.g. 5–10% of purchase cost)
— amortization period for cash
— labour rates for operators
— set-up, loading and unloading times for each tester
— programming times
— time to test a good board
— time to test a faulty board
— average time for fault diagnosis
— anticipated (or known) fault spectrum
— anticipated number of faults per board
— interface jig costs
— average number of iterations for a faulty board
— fault coverage for each type of tester
— availability of each tester
— average cost per repair cycle.

Most of the above points are perhaps self-explanatory, with the possible exception of:

Amortization. Clearly the way the company handles depreciation of test equipment influences the financial appraisal of ATE investments. The accountants are no doubt particularly interested in things like return on investment, or internal rate of return, payback period, and the like [13,14]. They will give greater weighting to short-term cash flows, so that returns in four or five years time will be valued less than income or outgoings near the start of the project. This they do by discounted cash flow (DCF) calculations. Not at all a difficult concept, DCF is simply the inverse of compound interest. For example, £1000 to be received in four years time would be equated to:

$$\frac{£1000}{(1.15)^4} = £438 \text{ at net present values (NPV)}$$

assuming in this case a discount rate of 15% per annum.

Availability. This is the percentage of time that a piece of equipment is available for use, either for programming (if this must be done on-line) or for testing boards. It is a function of reliability (MTBF) and the mean time to repair.

Fault coverage. There is no single figure for fault coverage of a test system. The coverage depends upon the type of circuit under test. Normally, a FNT would be expected to achieve around 90–95% while an ICT typically averages say 80–85%. Against this, FNTs are generally much more expensive than ICTs. However, some of the more advanced digital ICTs achieve over 95% fault coverage on typical digital boards. ICTs usually provide poorer fault coverage when analogue circuitry is being tested. It is therefore important to obtain information on anticipated fault coverage for boards of a type similar to those for which the facility is being planned.

Average Cost Per Repair Cycle. The ability of the diagnostic output of a tester to pinpoint a fault accurately significantly influences repair costs. If diagnosis is limited to indicating only the number of a faulty node of the circuit, then the repairer might replace several components before correctly identifying the faulty one. Thus the skill level (and hence annual cost) of the repair staff can be an influencing factor in an ATE investment appraisal.

Ultimately, one is dealing with a financial decision, and one could program a spread-

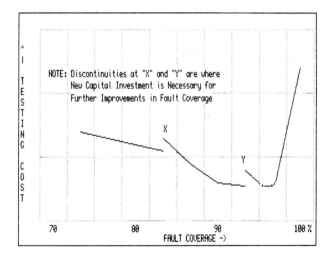

Fig. 6.10 Testing cost vs fault coverage

sheet with the above information. This is not by any means a trivial task, even though most good spreadsheets understand such terms as NPV.

SUMMARY

The aim of a test strategy is to minimize testing cost by separating good from faulty products at the most appropriate point in the product life cycle. The cost of finding a failure increases typically in decade steps as the point of interception moves from component level through board and system level. Faults occurring once the product has been delivered to the customer may carry a very severe financial penalty.

Component testing and pre-assembly 'burn-in' of active devices can help minimize expensive board re-work cycles due to faulty components and infant mortalities. Residual bad components plus faults introduced during manufacture and assembly of PCBs or hybrid microcircuits must then be captured at board test. In-circuit tests and / or functional tests need to be carefully planned in order to provide optimum fault coverage: complete fault coverage carries an infinite cost overhead.

REFERENCES

1. Bennets, R.G. (1984) *Design of Testable Logic Circuits*. Addison-Wesley, Wokingham.
2. Davies, B. *The Economics of Automatic Testing*. McGraw-Hill, New York.
3. O'Connor, P.D.T. (1981) *Practical Reliability Engineering*. Heyden, London.
4. Hnatek, E.R. (1977) Microprocessor device reliability. *Microprocessors and Microsystems*, **1**(5), 299–303.
5. Hawes, M.J. and Morgan, D.V. (Eds) (1981) *Reliability and Degradation — Semiconductor Devices and Circuits*. Wiley, New York.
6. Moralee, D. (1980) Economics of using ATE. *Electronics and Power*, **26**(2), 176–82.
7. Bennets, R.G. (1981) *Introduction to Digital Board Testing*. Edward Arnold, London.
8. Ost, G. (1985) The anatomy of burn-in — an analysis. *Electronic Production*, **14**(2).
9. Bullock, M. (June 1984) The impact of surface mount devices on in-circuit testing. *Proc. ATE East Conference*.
10. Smith, M. and Cook, S. (August 1985) Probe problems with SMDs. *Electronic Production*, **14**(8).
11. Poe, E.C. and Goodwin, J.C. (1981) The S-100 & other micro buses. 2nd Edn. Howard W Sams, Indianapolis.
12. Foley, E.G. and Firman, A.H. (1976) Testing microprocessor boards automatically. *Computer Design*, **15**(12).
13. Van Horne, J.C. (1977) *Financial Management and Policy*. Prentice-Hall, Englewood Cliffs, New Jersey.
14. Alfred, A.M. and Evans, J.B. (1971) *Appraisal of Investment Projects by Discounted Cash Flow*. 3rd Edn. Chapman and Hall, London.

7 Computer aided test systems and testability

7.1 OBJECTIVES

In this chapter typical systems for implementing the testing techniques outlined in Chapter 6 are considered. Equipment for testing components and assemblies is discussed by reference to examples of current products. An appendix to this chapter provides summary information on a range of automatic test equipments suitable for use in the production of electronic components, equipment and systems.

Also in this chapter the subject of testability is considered. In particular, aspects of design which significantly influence the testability of electronic components and assemblies are discussed at this stage, rather than in Chapter 2, since with the application of computer aided test techniques this subject must be considered in conjunction with the particular test philosophy and the means by which it will be implemented.

In reading this chapter you should gain an insight into:

— features of typical component testers
— features of typical in-circuit and functional PCB testers
— the concept of testability
— techniques of design for testability of VLSI chips and PCB assemblies.

7.2 TYPICAL AUTOMATIC TEST EQUIPMENT

As a means of illustrating how the techniques discussed in Chapter 6 can be realized in practice, some typical automatic test systems in current use are described in the paragraphs below.

7.2.1 Deltest 3300 component tester

Fig. 7.1 shows the 3300 from Deltest Systems Ltd. The 3300 is a versatile, low-cost tester capable of handling a wide range of components. Flexibility is obtained by means of interchangeable 'family modules' (Fig. 7.2) which plug into the test set. Modules are available for testing:

Fig. 7.1   Deltest 3300
Component tester

— linear active devices
— standard and custom digital devices
— diodes and transistors
— A–D and D–A converters
— triacs and thyristors

In addition, a general purpose family module is available which an OEM can adapt to his own requirements (e.g. for relays, timers, resistor networks, etc.)

For low-volume applications, devices can be plugged into and out of the tester manually. When large quantities of devices are to be tested, the 3300 can be interfaced with automatic component handling equipment. Also available as a software option is a statistical analysis package which acts as a data logging facility, provides print-outs of test results in summary or graphical form, and can be used to compare pre- and post burn-in test results. Percentage parameter drift can, for example, be analysed as a means of detecting devices with infant mortality characteristics.

7.2.2   Wayne Kerr Series 8510

The WKR-8510 series of low-cost ATEs (Fig. 7.3) provide single-pass combination testing facilities, whereby the test system can be used for continuity testing and in-circuit testing and functional testing of loaded PCBs.

Facilities are available for initially screening boards for short- or open-circuits prior to in-circuit testing. The system can check resistance, capacitance, inductance and total impedance between nodes as well as checking semiconductor junction operation. A module is available for tri-state impedance measurements.

A range of stimuli and sensors is available for functional testing of analogue and digital circuitry, and the system can be further extended, via an IEEE 488 control interface, to allow other programmable instruments to be incorporated into the system.

Digital

General
Purpose

Diode and
Transistor

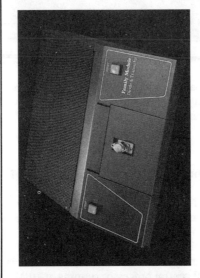

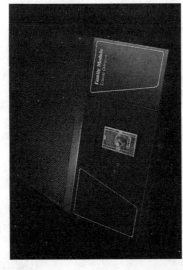

Linear

Thyristor
and Triac

Data
Converter

Fig. 7.2   Deltest 'family' modules

*Fig. 7.3 Wayne Kerr model
8510*

Loaded boards are held in place against the 'bed of nails' by means of a vacuum hold-down system (Fig. 7.4) which can be mounted within the test bed. Alternatively, an external vacuum supply can be connected.

7.2.3 Zehntel 850 analogue/digital ICT

The Zehntel 850 is a UNIX-based ICT for analogue/digital PCBs. Flexibility is obtained by means of a modular architecture offering a range of specialized analogue and digital in-circuit and functional test facilities. The 850 is pre-wired for 1024 universal test points and contains user-friendly software facilities for program generation, failure data analysis and data communications via Zehntel's NETCOMM. The system uses a 68 000 central processor running at 10 MHz clock rate, with a second processor,

*Fig. 7.4 Vacuum hold-down
system*

Intel 8086, within the testhead. Private memory of 1 MB and system memory of 512 KB are supplemented by Winchester disk and tape cartridge stores.

A feature of the Zehntel range is the availability of robot board handling systems (seen in Fig. 7.5 loading a board onto the tester). The Intelledex 605 robot locates a PCB from its handling box or conveyor system, and logs it into a computer listing via a bar code scanner. The robot then positions the board accurately onto the ATE fixture. An optional vision system is also available by means of which the robot can pre-screen the boards, rejecting those with missing components.

The test sequence commences automatically and upon completion the ATE passes a 'go/no go' signal to the robot which unloads the board from the test fixture and places it in the appropriate 'pass' or 'fail' bin. Via the computer logging system, a listing is automatically produced detailing which components have failed on every rejected board. The robot then picks up the next PCB and the cycle is repeated.

Via an Ethernet network and the Zehntel 700 net workstation, a multi-station test department network can be configured. This permits 'paperless' repair systems to be integrated into the test cycle, with rework information being routed automatically to the appropriate location. Historical records retained within the 700 net workstation can be used to generate reports on yields, product quality and inventories of work in progress.

Table A7.1 in the appendix to this chapter contains a summary of some of the key features of a selection of automatic test equipment in current use.

7.3 TESTABILITY

The previous discussion of automatic testing methods was general in its application and

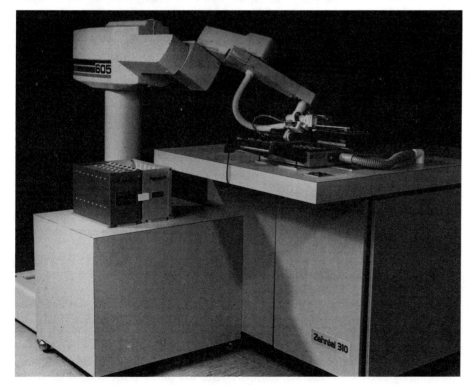

Fig. 7.5 Zehntel robotic PCB handler

fulfils the majority of requirements at the present time. However, as the scale of integration increases, the degree of difficulty in testing increases exponentially.

7.3.1 Testability of sequential logic

Combination logic can usually be partitioned into more testable 'chunks', but most vlsi circuits involve sequential logic for which such an easy solution is rarely available. Exhaustive functional testing of vlsi devices by conventional means becomes impracticable due not only to the speed limitations of FNTs but also the unacceptability of investing what could be several man-years in programming the tests. Automatic test pattern generation at the design stage can only go so far in alleviating these problems. Ultimately, as device and PCB complexity continue to rise, other measures need to be taken to ensure that production processes can be validated rapidly and good output delivered with high confidence. Here we look at just some of the techniques which are incorporated at the design stage in order to improve the testability of vlsi devices and PCB assemblies.

7.3.2 VLSI device testability

A vlsi device is likely to be testable if it has been designed with testability in mind: otherwise it is most likely to be untestable. The trend towards higher scales of integration has not been matched by a corresponding improvement in packaging capability. In 1975 a 1000-gate chip was considered very complex, yet the designer could specify a 24-lead package — around one pin per 40 gates. In 1985 a moderately complex device would contain 100 000 gates: where is the 2 500 pin package to provide a similar level of access to the internal circuitry? It is common today to work with more than 400 gates per package pin. Thus the internal circuitry becomes more deeply 'buried' within the chip, and this leads to a need for longer and longer test pattern sequences for these types of devices due to the reduced 'controllability' and 'observability' of internal gates.

Design for device testability

Several techniques have been devised to help achieve testable designs[1]. One of the most publicized is level sensitive scan design (LSSD), a technique devised by IBM in the late 1970s [2]. Various other scan-path techniques also exist and are all effectively aimed at improving the controllability and observability of a design [3].

Scan techniques involve chaining together the latches on a chip into one or more shift registers or 'scan paths'. Test data can then be shifted into the chip, clocked through the internal combinational logic circuitry, and then shifted back out of the chip via the scan path. The task of testing the chip has thus been reduced to simply testing the combinational logic, and since — subject only to pin-out limitations — one can increase the number of scan paths, it is possible to use the 'divide and conquer' philosophy to split up the combinal logic testing between several scan paths[4].

To use scan-path techniques, the latches on the chip must, of course, be capable of being switched between two modes of operation. Shift register latches (SRL) are master–slave flip-flops as in Fig. 7.6. In normal operation they behave as if the scan path were not present, while in test mode they operate as a scan path. Scan-path techniques carry an overhead in terms of silicon area or 'real estate', typically adding up to 15% area utilization by latches. In addition, normally three I/O pins are also committed to the

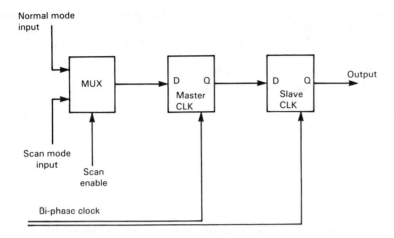

Fig. 7.6 Shift register latch

scan testing requirement, and there is more interconnection pattern on the chip which affects its 'routability'. Scan design can also carry penalties in terms of device performance in the normal operational mode, particularly in terms of speed. One reason for this is the increased set-up time necessary for the flip-flops due to the multiplexing stage.

It is inevitable that scan designs will carry a penalty in terms of reduced device yield, but at least it should be possible to eliminate most faulty devices in test using these techniques. An economic design philosophy may be to incorporate scan-testing facilities only for those areas of circuitry which are deeply 'buried'.

Scan techniques do not solve all problems of testability, however. For large memory chips, such as RAMs and ROMs, scan-path techniques prove difficult to apply. This is due to their serial access which implies a need for very long test patterns. Scan-path techniques also fail to detect a class of fault found to occur in CMOS technology, which is becoming an increasingly popular medium for vlsi implementation. It has been found that certain failure mechanisms can cause combinational logic to become sequential in behaviour [5]. Modified scan-path techniques have been devised, for example using two sets of latches within the scan path, to apply pairs of test patterns. Again a further silicon area overhead is inevitably involved.

While certainly not offering a total solution to the problems of vlsi testability, scan techniques can provide a route to testability at a cost far more acceptable than producing untestable devices.

Built-in self-test

At a further overhead cost in terms of chip area, it is possible to incorporate into the chip test equipment which can perform self-test at full functional speed. This can provide greatly increased cover of timing-related faults such as slow switching gates and hazards. One popular method of achieving built-in self-test (BIST) is by signature analysis [6] performed in conjunction with a test pattern generator. The test patterns can be read from a ROM or, more commonly, they may be a pseudo-random sequence generated by a linear feedback shift register. An n-bit register generates sequences containing $(2^n - 1)$ states. Again, it must be possible to multiplex between the test mode and the normal mode of operation.

Signature analysis is a method of compressing the response strings from a logic circuit into a single word — called the 'signature'. This signature can then be compared with the fault-free response of the circuit. There is a finite chance that a faulty circuit can give a fault-free response, but this probability is quite small when long sequences are used [7]. The fault-free response may be obtainable by simulation at the design stage, although the simulation run times needed may prove prohibitively expensive. Another way, assuming that some known good structures are available, is to actually log their signatures.

BIST is a trade-off between producibility and testability [8,9]. Signature analysis as a means of implementing BIST has the advantage that in many cases it can be accommodated during design without unduly degrading the primary performance characteristics of the device.

7.3.3   Board testability

Serious limitations exist with the traditional approach to board testing when one comes to extend it to complex vlsi-based PCBs. For ICT the limitations of test-pattern quality and quantity, of inadequate tester speeds, of the time necessary to execute sufficient test patterns, and the effects of long leads within the test fixture have been discussed. For FNT the unacceptably long programming time and limited diagnostic capability pose increasingly more difficult problems as integration scale advances. There are, however, things which can be done to offset some of these difficulties. Some design 'rules' for vlsi boards aimed at improving testability include:

— Provide good access — preferably via the edge connector — for important circuit nodes, address, data and control bus lines.
— Provide facilities for breaking feedback loops.
— Allow breaking of 'wired-ORs' for testing.
— Provide facilities for pre-setting of counters, latches and other memory type devices.
— Allow for disabling of free-running clocks.
— Avoid using multiple independent clocks on one PCB. The ATE can only synchronize to one clock at a time.
— Avoid mixing SIC technologies. If ECL is needed (for its speed) in part of the circuit, separate it physically from the MOS if possible and allow the ATE access to the ECL / MOS interface for testing.
— Return unused RESET and HOLD lines to V_{cc} or ground via resistors rather than directly. They can then be overdriven by the ATE if required.

Intuitively, one might expect that using an ICT with 90% fault coverage followed by an FNT with 95% coverage would leave very few faults undiscovered. Unfortunately, both techniques on their own tend to reveal the same sorts of faults, leaving a substantial residual defect level to cause problems at a later stage. ATEs which integrate ICT and FNT are being introduced as a means of improving test coverage. Starting with individual components or small areas, an ICT/FNT test combination is used to look for short-circuits and to check the component functionality. This process is repeated a number of times covering progressively larger but overlapping areas of the circuitry and culminating in FNT of the complete assembly.

7.4 INTEGRATION OF DESIGN AND TEST

The very strong links between design and testability of complex electronic components and assemblies is further justification for the integration of design and test. Considerable progress has already been made in the area of logic system design and test, with many of the more advanced logic simulators providing facilities for the generation and evaluation of test patterns. Such systems are applicable not only to designs in silicon, but also to PCB assemblies and to complete logic systems.

In a flexible manufacturing system, PCBs can be identified automatically — for example by a robot board handler reading a bar code on the PCB — and this information would be used to initiate the transfer of the appropriate test schedule from the central database into the ATE.

At present many elements of the test specifications for components and sub-assemblies of an equipment cannot be generated automatically, and must be inserted into the database by the design engineer. However, provided similar computer aided test systems are used during product development, the specifications so produced should be compatible with the production ATE and repair equipment.

SUMMARY

Last year's equipment fits on this year's PCB, and last year's PCB may now be a single chip: but the larger the scale of integration the greater the testing problem becomes. VSLI device testability needs to be designed in, otherwise circuitry may be deeply embedded within the chip where its controllability and observability may be very poor. Built-in self-test and scan-path techniques are used for improving the testability of vlsi devices. The testability of vlsi boards can likewise be improved by applying 'design rules' aimed at allowing the ATE closer control of the circuitry under test.

Some of the most advanced ATEs provide valuable automatic program generation facilities and provide detailed statistical analysis of results and trends. Integrating the 'test' function with the 'design' and 'manufacturing' database (Fig. 7.7) means that the data created during the simulation of the product can be used directly by the automatic test equipment, and fault diagnostic information may be linked directly to a repair station.

Robot board-handling and transport systems can automate the loading and unloading of components and PCBs onto and from automatic test stations. Similarly such systems

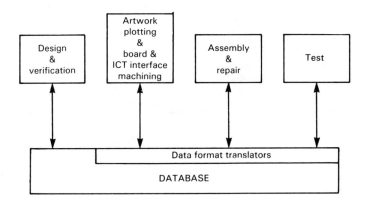

Fig. 7.7 Integrating ATE with CAD and CAM

can ensure that faulty boards arrive at the repair station where the necessary repair instructions from a central computer are available.

REFERENCES

1. Baillie, D. (March 1985) Design for testability: some practical considerations. *Silicon Design*, **2**(3).
2. Williams, T.W. and Parker, K.P. (1982) Design for testability — a survey. *IEEE Trans. Computers*, **C31**(1), 2–15.
3. Goldstein, L.H. (1979) Controllability/observability analysis for digital circuits. *IEEE Trans. Ccts and Systems*, **CAS-26**(9), 685–693.
4. Bennetts, R.G. and Anderson, J.R. (1985) Design for test solutions. *Silicon Design*, **2**(4).
5. Wadsack, R.L. (1978) Fault modelling and logic simulation of CMOS and MOS integrated circuits. *Bell Syst. Tech. J.*, **57** 1449–1474.
6. Hewlett-Packard (1977) A designer's guide to signature analysis. Application Note No. 222.
7. Bennetts, R.G. (1984) *Design of Testable Logic*. Addison-Wesley, Wokingham.
8. Konemann, B., Mucha, J. and Zwiehoff, G. (1979) Built-in logic block observation technique. Proc. IEEE Test Conf., 37–41.
9. Easterbrook, J. (1985) *BIST* — Built-in self test. *Silicon Design*, **2**(4).

APPENDIX A7.1

Some of the key features of a selection of current automatic test systems are compared in the following tables. It should be noted that the lists of existing ATEs are provided as representative of current technology, but constitute only a small sample of the total range currently available.

Table A7.1 Typical component testers

Supplier	Product	Application	Features
Analog Devices	LTS200	IC testing	Up to 600 V/20 A Floppy disk storage
GenRad	173X	IC testing	Magnetic tape storage Devices loaded into a range of 'family modules'
Fairchild	SENTRY 21	IC testing	High-speed (40 MHz) tests of gate arrays, memory and microprocessors, etc.
Hartley		Harness testing	Based on Olivetti M24 microcomputer Up to 4096 test points
Olivetti Technost	WPV68	Bare-board testing	Based upon Olivetti microcomputer
Teradyne	N151	Bare-board testing	Open- and short-circuits Bed-of-nails interface
Vanwall Datasystems	VDS8000	IC testing	D.C. and high-speed testing of LSI/VLSI devices

Table A7.2 Typical in-circuit testers

Supplier	Product	Application	Features
ATE Systems Computer Automation	BEAVER CUB 8100	Analogue/digital	1 MHz test rate Menu-driven control IEEE488 bus 20 μs/step
Eaton	Excel 407	Digital boards	Test rates to 20 MHz 256 test channels
Factron	730	VLSI boards	Test rate to 10 MHz + signature analysis 2048 channels
Marconi	ORION 2210	Bare/loaded PCBs Analogue/digital	1600 channels IEEE488 bus
Zehntel	300	Analogue/digital	600 channels IEEE488 networked

Table A7.3 Typical functional testers

Supplier	Product	Application	Features
Eaton	800	Microprocessor and LSI boards	IEEE488 bus Guided probe diagnostics Tests to 20 MHz
Factron	720	Analogue/digital	IEEE488 bus 30 MHz test rate Guided probe diagnostics
GenRad	1796	Analogue/digital	Automatic fault dictionary creation + guided prob diagnostics
Mars Electr's	GAZELLE FT	Analogue/digital	Self-learn modes IEEE bus Logic analyser modes
(Microcomputer can also run business applications software)			
Vanwall Datasystems	8000L	Analogue/digital	Tests to 5 MHz 256 test channels

Table A7.4 Combined functional and in-circuit testers

Supplier	Product	Application	Features
ATE Systems	ORAC	Analogue, digital and hybrid	Automatic program creation Automatic self-calibration Automatic acceptance limit tolerancing option
Factron	MB7700	Analogue/digital	Tests to 5 MHz 512 digital channels 224 analogue channels
Teradyne	L210	Analogue/digital	Uses DEC PDP11/44 CPU Tests to 10 MHz
Hewlett-Packard	306X series	Analogue/digital	Tests to 2 MHz Based on HP computers Up to 1048 channels
Factron	750	VLSI digital	30 MHz functional 10 MHz in-circuit Diagnostic software

Table A7.5 Computer aided repair systems

Supplier	Product	Applications	Features
Factron	40/404	PCB repair	Integrated into test system; repair instructions and required components are delivered automatically. Colour graphic display, quality control data collection and change documentation.
TEJAS	2200	PCB repair	Light beam identification of fault location
Zehntel	IRIS	PCB repair	(Integrated rework information system) Bar-coded PCB identification Light pen, cursor key and mouse as user interfaces Paperless history documentation database

8 Management systems in a computer aided electronic engineering environment

8.1 OBJECTIVES

In this chapter the problems of managing a business in an environment where design, manufacture, test and repair are achieved using computer-based tools are considered. The concepts and some of the realities of organizing an operation around a common product or project database are examined. Some of the computer-based tools for planning and control of electronic engineering projects are introduced together with other management information systems necessary for effective control and optimization of financial performance. In reading this chapter you should gain an insight into:

— product and project databases and their creation
— computer aided project planning systems
— management information systems
— management control systems which link technical and cost / time achievements
— decision-making techniques applied to investments in computer aided electronic engineering.

8.2 PRODUCT DATABASE CONCEPT

The product database, sometimes referred to as a 'design database', is a software system to assist in keeping control of the design of a product throughout its life cycle. In a dynamic electronic design environment there are continual changes in manufacturing processes, in component types and in the content of a design as it progresses through the design cycle. It is necessary to record and to control these changes so that upon completion of the design it is compatible with the manufacturing facilities, and components are available for assembly. A design database, with a suitable design management system (DMS), provides a means of achieving control of these aspects and also, as indicated in previous chapters, of the processes whereby data captured in the earliest design stages is transferred to subsequent stages.

8.2.1 Design management systems

Design management systems have been in existence for several years [1] and can provide improved security of data via structured levels of access and change authorization (Fig. 8.1). Thus, for example, the component data library for a particular project would be a sub-set of the company's central library of components. The project library might be made accessible to all members of the design team on a READ ONLY basis; additions to the library would only be authorized to those with EDIT authorization (e.g. project leader or technical administrator); and central library DELETIONS would require the authority of the Library Database/Administrator.

Once the early 'what if' stage of conceptual design has been completed, historical records of design changes are kept. Levels of authorization for changes are raised progressively as the design proceeds through its life cycle. At production release, the design data can be 'locked' so that change can only be introduced thereafter via agreed recording and authorization procedures. Such a system of configuration control makes for efficient use of time and storage space, since lower issue 'documents' are not maintained in total. It is only necessary to record the detail changes at each revision, archival copies being re-created if required in the future by a 'walk-back' process.

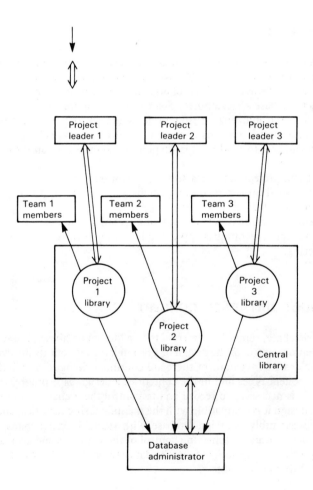

*Fig. 8.1 Library/database
access and control*

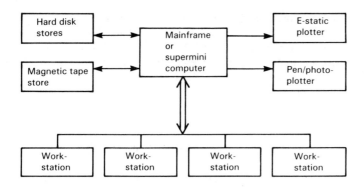

Fig. 8.2   Typical design facility

8.2.2   Control of design activities

The design management system can be used to record and control design activities. To achieve this, the DMS must therefore reside in the data route between the design data-base and the various software design tools. Thus when a schematics editor is used for initial data capture, the design management system might typically perform data validity checks (e.g. seeking net list errors, unnamed components, etc.) before releasing the design to the database. Simulation of the circuit involves transfer of the circuit data, with appropriate format conversion via the design management system if necessary, to the logic simulator. The design management system must be compatible with the range of physical design software tools which will be used for hardware realization (e.g. PCB, hybrid microcircuit, gate-array, etc.)

Figure 8.2 illustrates a typical design facility consisting of microcomputer stations for schematics capture, workstations for simulation and modelling, dedicated PCB design systems and a minicomputer for system management and central library functions. In this example the central computer also provides reporting and plotting facilities.

8.2.3   Decisions in back-up and security

By the use of appropriate data translators, where necessary, the design data stored within the product database can be transferred between design and manufacturing tools, even though some of these tools may be located remotely — perhaps owned by another organization. Each transfer must be routed via and controlled by the database manage-ment system, and problems can arise when it is required to transfer data between com-panies. There is no single agreed standard for these data. Thus if a company wishes to subcontract PCB machining or even layout work modifications (as might be necessary in the event of major damage to the company's own facilities) then it would be necessary to select a subcontractor with compatible CAE systems.

It is sound management practice to establish these sorts of mutual agreements so that the business can survive a fire in its CAD office, for example. Having a back-up set of data tapes is of little help if there is no hardware system available to run it.

IGES

For two-dimensional and three-dimensional (wire-frame) geometric data a graphics standard is in existence. Initial graphics exchange specification (IGES) has been developed on US military contracts [2] and provides for general entities such as lines and curves and certain types of surface patches. Unfortunately, IGES does not, at the time of writing, support schematic diagram representation of electronic circuits. Considerable interest is being shown in the subject of CAE standards, as these will greatly assist the achievement of system integration.

8.3 PROJECT PLANNING, MONITORING AND CONTROL

The computer-based tools considered in the foregoing chapters are aimed at assisting in various design and manufacturing activities. It is crucial to the successful completion of a project that the start dates and completion dates of these activities are compatible. Clearly, design tasks and manufacturing tasks are interdependent in terms of technical content. Timing must also be phased so that the necessary information for a manufacturing process is available when the other resources (manpower, materials and facilities) require it. For the planning of complex design and manufacturing processes, various techniques have been devised and computer aids for their implementation are available to management.

8.3.1 Bar chart planning

The simplest forms of planning aids consist of programs which convert numerical data detailing start and finish dates for important project tasks (called 'work packages') into graphical displays (Fig. 8.3). The charts provide a manager with rapid access to information on estimated timescales for major activities across the project life cycle. The decisions on required start dates for work packages must be determined by the planner himself using his knowledge of the dependencies of work packages upon one another. For relatively small projects carried out in an environment independent of other projects, bar chart planning programs can be an effective means of creating project plans. Progress is monitored by means of recording completion of 'milestones' which

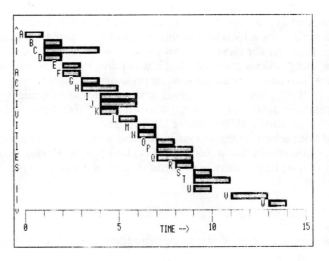

Fig. 8.3   Bar chart report format

are major events within the work packages. Usually the completion of a significant action or delivery of a product or service to the customer is used as a project milestone.

8.3.2 PERT

Project (or programme) evaluation and review technique (PERT) is based upon 'activity on arrow' networks the structure of which not only reflects the resource implications of each activity, but also its dependency upon the other activities within the project [3]. Fig. 8.4 illustrates a typical format for representation of a single activity.

The input to a PERT planning package is a list of activities; estimates of the minimum duration and resource requirements; and start and finish dependencies for each. The program predicts earliest and latest start dates and also earliest and latest completion dates for each activity. If a change in the duration of an activity results in a similar change in project duration, the activity is said to be critical. A path, termed the 'critical path', can be traced through the network from start to finish following these critical activities. Management resources would then be concentrated on the progressing of this critical path.

On some PERT packages resources may be categorized in general form (e.g. software design engineer, full-time for seven weeks) or allocated as individuals (e.g. 25% of J. Smith's time during a particular four-week period). PERT programs generally provide facilities which attempt to optimize such project parameters as achieving minimum timescale or working within a fixed resource budget. When timescales are compressed, which is almost always the case on electronic engineering projects, it is likely that an initial plan will contain demands for resources which fluctuate considerably in level. In practice this could necessitate the availability of an over-provision of resource for much of the project duration. However, activities not on the critical path contain some 'slack' in respect of start times, and improved resource loading can be achieved by moving the start dates of non-critical activities. On all but the most trivial of projects, attempting to achieve this level of planning sophistication by manual means would present a very daunting prospect. When several projects compete for resources within a department, multiple-project planning systems are needed.

```
10                     Environmental tests, Module 4
                                  7.5
       EXAMPLE:    [ 6 ]=================================>[ 7 ]
                 (2:Jan, 5:Feb)  Eng 2.0   Techn 7.5   (23:Feb, 28:Mar)


                       Activity description / code
                            Planned duration
 5     KEY:       [ I ]=================================>[ J ]
                         Resource code(s) & estimates
                 ( ES / LS )                      ( EF / LF )


                 I = Start event        J = Finish event

                 ES = Earliest start date   EF = Earliest finish date

                 LS = Latest start date     LF = Latest finish date
 0
```

Fig. 8.4 PERT representation of an activity

8.3.3 Project monitoring and control

A pre-requisite to effective control is effective monitoring. An effective monitoring system must compare actual performance with planned performance in terms of the complex interrelationships between time and cost. This involves measuring the value of work done, from which forecasts of total cost and timescale to completion can be derived.

8.3.4 Line of balance

In the manufacturing stage of a project it is usually not too difficult to measure the quantity of output where finished goods are concerned. It may be a little more complex, however, to monitor progress if a substantial amount of work is in progress through the various manufacturing processes, and here 'line of balance' (LOB) techniques are available [4].

In manufacturing one is dealing with repetitive operations with timescales and resource demands which generally contain less uncertainty than for design activities (except perhaps at production start-up or when facilities are changed, at which times the effects of 'learning' may distort the pattern). The network for producing a single item may typically contain less than 50 processes or activities, the completion times of which can be determined (e.g. from a simple PERT network) and hence the 'lead time' for each milestone can be measured.

A simple example is outlined in Table 8.1. The network for building a single controller is shown in Fig. 8.5, and the schedule of customer deliveries is shown in Fig. 8.6. The programme bar chart (Fig. 8.7) shows the quantity of actual deliveries on the vertical axis. The horizontal axis contains elements corresponding to each of the milestones. The LOB indicates the cumulative numbers of parts, sets of components, etc., for each milestone which must have been completed at the date of progress review in order to be on target for completion of planned deliveries at the schedule date. The LOB is thus derived by measuring forwards along the schedule timescale from time 'NOW' (the review date) by an amount corresponding to the lead-time for each milestone. The schedule deliveries at that point coincide with the number of milestone items which must have been completed at the date of review.

The number of 'actual' completions for each milestone are recorded on the programme chart so that discrepancies from the LOB are immediately apparent. Points

Table 8.1 Assembly of energy management system controller

Milestone	Description	Lead-time (weeks)
1	Receive enclosure parts	5.0
2	Begin PCB manufacture	4.4
3	Begin enclosure manufacture	4.7
4	Receive electronic components	3.9
5	Complete enclosure assembly	2.9
6	Begin PCB assembly	3.6
7	Receive front-panel controls	2.0
8	Complete PCB assembly	1.5
9	Test and install PCB	1.5
10	Connect front panel controls	1.0
11	Test system	0.6
12	Deliver to customer	0.0

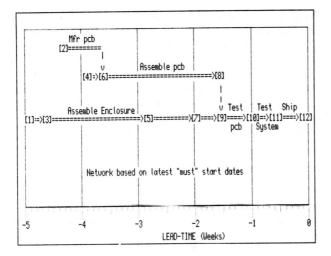

Fig. 8.5 Network for building one energy management controller

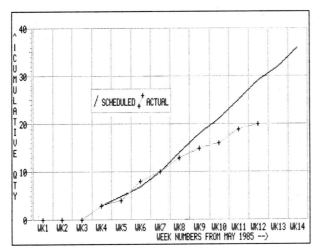

Fig. 8.6 Schedule and actual deliveries

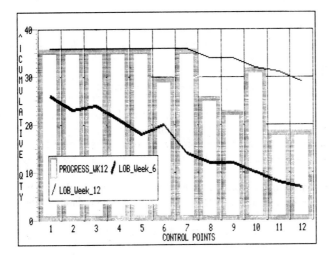

Fig. 8.7 LOB chart

below the LOB indicate that the project is behind schedule. (A 'culprit' milestone may, of course, result in failure to meet the LOB target for other milestones further 'downstream' in the manufacturing process.) Similarly, actual results well in excess of the LOB are indicative of 'work in progress' in excess of schedule requirements.

Where very large manufacturing projects — or many smaller ones — are involved, LOB schedules may be 'nested' hierarchically. From each sub-assembly one or two key milestones are selected for incorporation in the next higher assembly level. The highest level LOB chart therefore records progress on the project as a whole. Any shortfalls at the highest level may be further investigated at the next level down.

8.4 PROJECT MANAGEMENT INFORMATION SYSTEMS

The progress of a project, especially during development, requires careful monitoring so that corrective action can be taken to optimize performance against targets. Critical path analysis (CPA) of the PERT network serves to focus management attention upon the aspects of the project which most influence timescale achievement. Analysis of planned and actual progress and costs provide the basis for measurement of 'earned value' on a project as well as providing forecasts of resource requirements. (Resource requirements invariably deviate from the original or baseline plan, since it is by re-allocation of resources between tasks at the lowest levels of project activity that control is exerted to maintain major milestone targets.)

Project management reporting systems are available via which detailed information may be extracted by analysis of planned and actual performance data. Microcomputer software packages are available [5] which provide bar charts for progress monitoring, status reports on activities (whether critical or amount of slack); resource distribution histograms; and forecast and actual cost profiles.

More comprehensive management information systems use the project database concept [6]. At the initial planning stage a PERT plan is prepared and optimized in terms of skills resource levelling or spend profile. As the work proceeds, actual start/finish times and materials costs are inserted and the effect of any variances upon the project milestones is analysed by the computer. The manager may use 'what if' techniques to evaluate the relative merits of various control measures before selecting the best course of action. (In some systems a 'what now?' interrogation of the system may be used to obtain a re-plan via the computer program.)

Reports are available on activity status showing 'can start' or 'must start' dates; expenditure profiles based upon earliest and latest dates; earned value reports by work package and for the whole project; and budget, actual and forecast cost profiles. Figure 8.8 illustrates the latter report in graphical form.

Input to these types of systems comes from two sources:

(a) The records of labour committed against each work package. (This is normally derived from an analysis of weekly timesheets.)
(b) An assessment of percentage completion of each task or work package. (This is provided by the manager of the work package on a weekly or monthly basis.)

When both the amount of work completed and the spend to date differ from the plan, processing of the budget and actual cost and performance data can be used to provide forecasts to project completion. In Fig. 8.8 the implied programme slip, cost variance and schedule variance at project completion have been determined and are illustrated graphically.

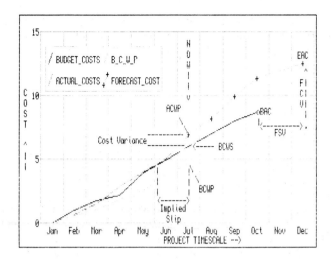

Fig. 8.8 Earned value analysis chart

Running on powerful mini- or mainframe computers, multi-project management information systems can also provide the necessary information for forward planning of investment in facilities, staff recruitment and training.

8.5 INTEGRATING TECHNICAL AND PROGRAMME SYSTEMS

A great deal of effort is now being directed towards the achievement of integrated systems for relating the technical decisions on a project to the financial and schedule implications. Thus the partitioning of a complex circuit into two PCBs where the original plan had assumed a single board would automatically reflect into an updated works cost (cost of manufacture) estimate, a revision of the estimates for physical design of the PCB layout, and subsequent assembly and testing phases both in development and in production. It is clear that such a design change also has implications in terms of size, weight, reliability and other parameters which ultimately may have an effect on sales order intake of the business in future years. The main point to note is that there is no such thing as a 'purely technical decision' in any industry, and the electronics industry is no exception. The linking together of all these systems via a common database offers the prospect of further improvements in business efficiency and hence in competitiveness.

8.6 CAE INVESTMENT PLANNING AND IMPLEMENTATION

Although there may be great disparity between the level of investment in CAE which would provide optimum results and the financial resources available for that investment, the major problem facing potential users of CAE systems is that there are currently no completely integrated CAE systems available for purchase. It has been suggested that suppliers of CAE systems did not recognize the value of integration early enough, and their systems are aimed at competing against the capability of other vendors rather than providing close compliance with the requirements of the end user [1]. As a consequence, the first-time user is faced with the very difficult task of making sound investment appraisal decisions in an area of technology which may be quite new to him, and of planning the integration of a range of complex tools in such a way that his existing and future operations can function efficiently during and after the implementation of CAE.

8.6.1 Communication

Integration implies communication. CAE systems built around a mainframe computer serving many users in an on-line mode clearly offer the advantage of a simplified route to further system growth. The dedicated workstation approach, in contrast, offers advantages in the degree of specialization available. However, once the required man–machine interface features are available at a workstation, the question of whether it is driven from its own internal CAE engine or obtains its performance via a mainframe is of little significance to the user. The capability is ultimately determined by the CAE software, and in particular the database management system. CAE workstation suppliers have responded to the need for integration — where a single common database is accessed by several users — by providing a networking capability (the Apollo Domain is one such example). Workstation users may thereby share data and other resources and can communicate with other CAE tools.

8.6.2 Planning timescales

In an industry where product life cycle times are continually shortening, there is a temptation to try to set similar timescales for the implementation and pay-back of capital investments. In larger electronics companies this pressure is further reinforced by the need of a divisional or business managing director to show real achievements in financial terms within a very short timescale. (In many organizations it is not uncommon for movement between posts at business director level to occur every two to four years.) Large-scale investment in CAE involves planning over timescales of several years, with phased benefits anticipated both in the medium and long term. Many electronics companies now measure the performance of the technical and general management at least in part by their ability to keep up to date with and to plan the application of technological advances in such areas as CAE.

8.6.3 Analysis of requirements

In selecting a route for implementation of CAE, it is essential to create an objective specification of requirements, both in terms of initial implementation and growth capability of the system. An analysis of the nature and level of current and planned business activity is the basis for definition of the technical requirements of the system. It is not unreasonable to phase the investment, in line with budget limitations, such that areas offering greatest short-term returns are implemented first — *provided* that in so doing the route to implementation of the complete system is not impaired. For example, the short-term benefits of a particular stand-alone CAD system for PCB design might appear attractive in the context of present requirements but if it is corporate strategy to make use of ASICs in future products, then selecting a CAD system which has no growth potential in this direction could be doubtful policy in the longer term.

Factors to consider in this context include:

— Does the supplier offer a growth path to higher levels of integration?
— Are links to other CAE tools available (e.g. data format translators to match the requirements of photoplotters, automatic assembly systems etc.)?
— Can all tools operate via a central database?
— What is the suppliers 'track record' in terms of provision of an upgrade path for purchasers of his earlier generations of CAE tools?

8.6.4 Selection processes

There are currently several hundred suppliers of CAE equipment, and it is impracticable to carry out a complete financial appraisal on each of the options available. Some process is therefore needed for selection of a 'short-list'. One decision-making process is summarized in Table 8.2 where the factors important to the user are listed and ranked in order of importance. Some features will be considered absolutely essential, and systems not capable of providing them need not be considered further: they are not entered into the table. Each feature is then given a weighting factor related to its value to the purchaser. This weighting process can be rather difficult, but if one factor can be quantified — related to a financial benefit — then the notional value of other features can be related to cost by asking 'How much is it worth to us in financial terms to have a system which has optimum characteristics in respect of this feature?'

In the example of Table 8.2, three systems are compared and it is clear that options A and B are preferable to option C for the particular circumstances under consideration. It would not be reasonable, however, to choose between A and B without a more detailed financial analysis of costs and benefits.

8.6.5 Financial appraisal

The most commonly used method of assessing the value of an investment is to determine how long it will take before the financial benefits from the investment provide 'payback' of the initial investment. Thereafter there will be a 'return' on the investment.

The payback period is simply calculated from:

Table 8.2 Initial selection chart — automatic component testers

	Decision factor	Weighting	Weighted ratings of options System A	System B	System C
1	Versatility	10	9	8	7
2	Throughout	7	5	7	3
3	Diagnostics	4	2	3	3
4	Stat's analysis	5	3	4	2
5	Upgrade facil's	3	3	2	3
6	Delivery	4	4	3	2
7	Maint. support	6	4	4	2
	Sum of factors 1–7	39	30	31	22
X:	% optimum (effectiveness):		77	79	56
	Purchase cost		24 000	36 500	28 000
	Annual running cost		15 500	12 000	13 500
Y:	cost/yr. (Assuming 5-yr life)		20 300	19 300	19 100
	Cost/effectiveness (Y/X):		264	243	339

$$Payback = \frac{net\ investment}{average\ annual\ cash\ flow}$$

where annual cash flow is calculated from the cost savings or income from the invest-ment less any cost outgoings associated with it (e.g. running and maintenance costs).

Payback takes no account of the actual pattern of cash flows, however, so if in our example option A shows earlier returns on investment, and hence provides cash for con-tinued investment at an earlier date, payback analysis would not show this advantage. Equally, if option B is showing a rising return at the end of the payback period, would this not suggest that it is likely to continue to provide benefits for some time after the payback date? This, too, should be a factor considered in choosing between the projects.

8.6.6 Discounted cash flow

The very existence of interest rates in excess of general inflation suggests that money has a value which is dependent upon the point in time at which it is to be payed or received [7,8]. Discounting is a means of determining the present value of a sum of money to be received at a future date.

To illustrate this concept, consider a sum of £100 invested at 10% interest, com-pounded annually. The future value of such an investment is given by:

$$FV = PV(1 + r)^n$$

where FV = future value
       PV = present value
       r = interest rate (0.1 per annum in this example)
       n = number of periods (years in this case)

Thus after two years the future value is:

$$FV = £100(1 + 0.1)^2 = £121$$

If, on the other hand, a return of £100 is anticipated at two years time, then we can use the inverse process — called discounting — to determine the present value as:

$$PV = FV(1 + r)^{-n}$$

which gives a PV of 100/1.21 = £82.60 for the example above.

One can now set up a spreadsheet to calculate the net present value of investment options A and B from forecasts of expenditure and income over the anticipated lifetime of the projects. The analysis is illustrated in Table 8.3. An advantage of using computer aids to investment appraisal is that one can assess various 'what if' situations. In recent years interest rates have fluctuated widely, and it is far from clear what discount rate is likely to prove realistic over a project lifetime of several years. In this simple example the discount rate has been kept constant over the whole of the time period, but analysis has been performed with three different discount rates. The return on investment results are seen to be quite sensitive to interest rate fluctuations, with the relative financial advantage shifting from option A to option B should interest rates rise above 10%.

A more realistic financial modelling process might be to set discount rate at present-day values at the beginning of the investment period, and to adjust the rate upwards and downwards from that point in order to appraise the effects upon project viability.

One final point should be made concerning the use of financial appraisal modelling systems. The data upon which the calculations are made is determined by forecasting trends within the market. This involves assumptions about the economy in general,

Table 8.3 Comparison of projects by discounted cash flow and NPV rate of return on investment (N.B. values in £K)

	Project A				Project B			
Year	Investment1	Return1	Net Retn1		Investment2	Return2	Net Retn2	
1	£10.00	£0.00	£ −10.00		£25.00	£0.00	£ −25.00	
2	£25.00	£0.00	£ −25.00		£45.00	£5.00	£ −40.00	
3	£30.00	£0.00	£ −30.00		£10.00	£65.00	£55.00	
4	£7.00	£10.00	£3.00		£2.00	£50.00	£48.00	
5	£2.00	£15.00	£13.00		£2.00	£10.00	£8.00	
6	£2.00	£35.00	£33.00		£0.00	£5.00	£3.00	
7	£2.00	£40.00	£38.00		£0.00	£5.00	£5.00	
8	£2.00	£35.00	£33.00		£0.00	£0.00	£0.00	
9	£2.00	£25.00	£23.00		£0.00	£0.00	£0.00	
10	£2.00	£5.00	£3.00		£0.00	£0.00	£0.00	
Discount rate				% Retn inv.				% Retn inv.
5 %. NPVs:	£75.84	£123.27	£47.43	62.5	£81.87	£122.79	£40.92	50.0
10 %. NPVs:	£69.32	£93.81	£24.49	35.3	£78.28	£108.59	£30.30	38.7
20 %. NPVs:	£59.57	£57.03	£ −2.53	−4.3	£72.37	£86.75	£14.38	19.9

about customers and about competitors. It is all too easy to ascribe spurious accuracy to the data within an investment appraisal. Such techniques should be seen as decision aids to be used in support of innovative or entrepreneurial management, and not as a substitute for creative marketing and leadership.

8.7 PEOPLE AND CAE

Experience has shown that implementation of a CAE project requires careful planning and management. So far, technical considerations have been considered in some detail, but people are important, too. The impact upon and required contributions from staff should be considered when establishing an implementation plan, and hence the following points are worthy of consideration when planning a CAE project:

— The best CAE tools will only approach their full performance potential if the staff who use them are also committed to making a success of the project. It is important, therefore, that staff are involved in the decision processes. Key members of staff could, to advantage, take part in visits to appraise vendor equipments, or in meetings where the financial analysis is discussed. In many cases the investment in their time will be repaid several times over in the form of a smoother transition to the new system.

— It is easier not to change methods of working. Change involves hard work, and people need to see personal benefits if they are to be fully motivated to work towards the implementation goals. The implications of the proposed system, in terms of company performance and the impact upon job opportunities and responsibilities, are best discussed at an early stage. Fear of computers is fast receding as people see that the alternatives are 'automatic or liquidate', and that successful implementation of CAE gives best prospects of company growth. This should be understood as offering the only route to security of employment in an industry which by its very nature is committed to ever-increasing output per employee.

— Implementation of a CAE project is not something which can be accomplished by a busy manager using the odd few spare moments in the day. It requires a 'champion' [9,10] who has the full backing of senior management; the authority and management skill to establish targets and to set up a monitoring system so that their attainment can be verified; and sufficient expertise in CAE to be able to set up meaningful 'benchmark' tests of systems during evaluation, and post-delivery acceptance testing.

— Staff training in the use of complex CAE tools is required. It may take several months for operators to reach full efficiency on a new system [9]. Training completed prior to installation could be a means of securing a smoother transition from the old system to the new.

SUMMARY

Product or design database systems provide a means of recording and controlling the technical implementation of product development and manufacture. The database management system controls transfer of design data from one computer-based tool or process to another via appropriate data format translators. Standardization of data formats is an important consideration in the integration of design and manufacturing via a common database.

Project planning systems based upon time/cost analyses of large numbers of dependent activities are accomplished rapidly, and hence more thoroughly, by the use of computer programs. Techniques such as PERT and CPA allow a manager to evaluate options both at the initial planning and estimating stage and also as the means of controlling the project during implementation. Reports from these types of management information systems can be used for resource planning for a project, a department and the business as a whole.

Repetitive tasks, whether model-build programmes within a development project or in subsequent production, can be planned using LOB techniques. When many interacting projects are concurrent, hierarchical structures of PERT and LOB plans are created and maintained on computer. The next stage in system integration involves interactive linking of programme time and cost information systems with the engineering database.

Planning for the implementation of a CAE project involves making decisions on technical, financial and human factors. There is no longer a choice of whether or not to invest in CAE in the electronics industry. Increasingly it is the case that the work cannot be completed by any other means. However, the effect of a poorly planned and implemented CAE project could be just as serious for a business as a decision to postpone a 'decision'.

REFERENCES

1. Scroggins P. (1983) Integrating electronic design and management. *New Electronics*, 35–9 (September).
2. *Initial Graphics Exchange Specification (IGES) Version 2.0.* National Bureau of Standards, US Dept of Commerce, Washington DC, NBSIR 82-2631(AF), February 1983.
3. Lester, A. (1982) *Project Planning and Control*. Butterworths, London.
4. Schoderbeck, P.P. and Digman, L.A. (Sept./Oct. 1967) Third generation PERT/LOB. *Harvard Business Review*.
5. Examples of microcomputer software packages for project planning include the *PMS-II* suite from North America Mica, *PERTMASTER 2000, HORNET 4000,* and *PROJECT SCHEDULER 5000* from Scitor Corporation.
6. SDS2, a system developed by Software Sciences, is an example of a network database orientated project planning system suitable for very large-scale project management. Currently available on DEC VAX machines.
7. Van Horne, J.C. (1977) *Financial Management and policy*. Prentice-hall, Englewood Cliffs, New Jersey.
8. Alfred, A.M. and Evans, J.B. (1971) *Appraisal of Investment Projects by Discounted Cash Flow*. 3rd Edn. Chapman and Hall, London.
9. Green, A.J. (1983) The Management of a CAE Project, *Electronics and Power* (February).
10. Schofield, A.S. (1985) Strategies for acquiring and building integrated CAE systems. *Computer-Aided Engineering Journal*, 24–9 (February).

9 Emerging technologies and CAE

9.1 OBJECTIVES

In this chapter some significant trends in electronics and computer technology are outlined and their likely impact upon CAE is briefly considered.

This survey is, of necessity, both narrow in scope and limited in depth, but it should provide the reader with a pointer towards further study of developments as they emerge in the fields of:

— new electronic devices
— computers and communications
— CAE systems and factory automation.

9.2 NEW ELECTRONIC COMPONENTS

For nearly 40 years silicon has been the material used in the majority of active devices, and considerable investment is continuing with the aim of further improving the performance of silicon devices. In the USA the government is sponsoring programmes for the development of very high-speed integrated circuits (VHSICs) and, while the initial applications are military, this programme will also make available improved manufacturing technologies for industrial and consumer standard products and ASICs.

9.2.1 Silicon integrated circuits

In 1985, 256 Kbit dynamic RAMs became the standard building 'brick' for computer memories, while the first 1 Mbit RAMs were being delivered to OEMs. By the end of 1986 the 1 M RAM will probably have replaced the 256 K RAM for new designs, and the 4 M RAM, already produced in laboratory conditions in 1984, may be available on a sample basis. These types of devices use one transistor per memory cell, and hence a 1 Mbit RAM contains just over a million transistors (allowing for address decoding and other peripheral circuitry). The regular internal structure of memory devices considerably eases the task of physical design.

As device geometries are reduced in response to the pressure for acceptable yield and increased speed of VLSI devices, the demands on CAE systems for design, manufacture and test increase also. This is already posing a major problem in testing these VLSI devices [1]. Two techniques which are currently receiving a great deal of attention are:

Fault-tolerant circuits [2]

Yields of some of the most complex semiconductor devices are very low — below 10% in the early stages of production — and this results in a very substantial cost premium. Fault tolerance involves the incorporation of an amount of additional circuitry, into devices or onto PCB assemblies, which under normal circumstances has no effect upon the operation of the system. In the event of failure of the primary circuitry the 'redundant' circuits take over and maintain operation of the system.

Wafer-scale integration

If it were not for the very high probability of faults in one or more individual chips on a wafer, then long before now complete computer memories would certainly have been produced on a single wafer. Instead, the wafer is scribed and cut into individual chips which, after assembly and testing are re-connected via an expensive PCB. The multiple connections on each circuit path (chip to header, leads to PCB, etc.) consume space and degrade reliability. Wafer-scale integration is an emerging technology aimed at overcoming the yield difficulties so that all of the 'good' chips on a wafer can be used for a system application without separation of individual devices. The main problem is to identify the good devices and arrange for them to be connected to one another but not to be influenced by the faulty devices.

One interesting technique involves a system configuration procedure each time the circuit is powered up. On application of the supply, one of the 'good' chips acts as 'controller' and sends out messages over a bus to which all chips are connected. The other chips carry out self-tests and respond individually if they are working properly. The functions of each chip are then allocated by the controller and the system is ready for use.

9.2.2 Gallium arsenide integrated circuits

Since 1974, when The Plessey Company produced the world's first gallium arsenide (GaAs) monolithic microwave integrated circuit, the difficulties of manufacturing devices in this technology have gradually been overcome and costs are now reducing to the point where products other than the most crucial military systems can be considered as candidates for this new technology.

The main reasons for the intense interest which is being shown in GaAs stem from two important characteristics. First, the electron mobility in GaAs is up to six times higher than the mobility in silicon: this results in approximately three times greater operating speeds for the same device dimensions, or a similar speed to silicon but with between six and ten times less power consumption. The second feature of GaAs which distinguishes it from silicon is that it is a direct band-gap semiconductor, and can be made to emit light. Thus GaAs finds application in light emitting diodes (LEDs) and lasers.

The use of GaAs-based heterostructures (structures involving junctions between different types of semiconductors such as GaAlAs) provides gates with propagation delay times between 10 and 20 ps. Using these techniques, field effect and bipolar integrated circuits have been produced by several manufacturers. GaAs devices currently available include 1024 bit RAMs, multi-stage counters and dividers operating to 10 GHz, and low-noise amplifiers covering 2–30 GHz. In April 1985 the MEDL GaAs foundry service came on-line offering a gate-array service to OEMs. Similar facilities are being introduced by many of the seventy or so companies worldwide who are investing in this technology [3].

The electrical performance benefits of GaAs are the main reason for adoption of this much more costly fabrication medium. However, in certain applications the ability to survive higher operating temperatures and the greater hardness to radiation (For certain types of GaAs junction FETs the improvement compared with silicon is two or more orders of magnitude [4]) are also important factors.

9.3   COMPUTERS AND COMMUNICATION

Until the early 1980s, stored program computers continued to use architecture which differed little in concept from the machines devised by Von Neumann and his co-workers half a century ago. In the last two or three years devices aimed at parallel processing of data have begun to emerge with the objective of increasing computing power despite the speed limitations of electronic device technology. One example of this philosophy is the transputer developed by INMOS [5]. In principle any number of transputer chips can be used to create a computer having the required processing power for a particular application. It may be possible, using this type of technology, to build super-computers capable of processing data from seismographic surveys, weather forecasting and similar applications essentially in real time.

The benefits within CAE are likely to include the ability to test more circuits at their operational switching speeds, and more meaningful mathematical modelling of individual components. This latter point is particularly important when a simulation is to be performed using a combination of physical hardware for the established parts of a system and computer modelling for parts being developed or modified.

9.3.1   Artificial intelligence

Many of the problems encountered by workers in the field of artificial intelligence would largely be overcome if the power of computing engines could be raised substantially. For real-time vision and speech systems computers with greatly improved processing power and very large memories with fast access are urgently required. Individually the technological problems associated with each of these aspects have been solved. VLSI silicon chips with 10^7 gates are a realistic future prospect for the end of the present decade; sub-micron geometries in silicon and GaAs will provide fast-access memory devices capable of supporting the increased processor performance; and parallel architectures may emerge as effective solutions to the problem of continuous demands for increased processing power [6].

9.3.2 Communication networks and data transfer standards

Inevitably, as new concepts emerge, manufacturers vie for business by developing proprietary solutions with features which maximize their individual product offers. This is particularly noticeable in the field of network communications. Standards are now emerging, however: for example many CAE systems now offer Ethernet facilities via which several workstations for design, test and repair can be data-linked together whilst spatially separated [7].

Several IC vendors have developed their own data formats, so that translation between workstations of different types and between different fabrication processes involves the creation of numerous translation programs. Recently a number of companies, including Texas Instruments, Daisy Systems and Mentor Graphics have consolidated their standards by created the Electronic Design Interchange Format (EDIF).

EDIF is intended as a machine-to-machine interchange format and has a syntax similar to the LISP language used in some artificial intelligence applications. EDIF provides a format for the description of network structures, logical functions and physical layout shapes [8].

9.4 CAE SYSTEMS AND FACTORY AUTOMATION

The level of sophistication of electronic devices and equipments has now reached the point where it is quite impossible to produce state-of-the-art products without the use of CAE tools [9]. As complexity continues to rise it will cease to be acceptable to transfer data between the various tools by any means other than automatically. The truly integrated CAE system will extend beyond design through not only the main manufacturing processes but, more significantly perhaps, to the control of the design / manufacturing system itself: into areas of decisions previously considered to be the prerogative of 'management'.

9.4.1 CAE systems

The most significant development in the area of CAE tools will perhaps be the emergence of silicon (and GaAs) compilers. These systems will enable an engineer who is not necessarily skilled in device fabrication technology or circuit design to create manufacturing data for the realization, in semiconductor form, of an electronic circuit from instructions consisting of a high-level description of its function. The present-day silicon compilers discussed in Chapter 2 represent an important step in this direction.

A logical step onwards from the silicon compiler is the concept of an 'electronics product synthesizer' which would convert a requirement into the most appropriate form of realization. Such systems would thus provide physical devices or PCBs in response to a functional description of a required circuit.

9.4.2 The 'factory of the future'

Computer Integrated Manufacturing (CIM) has been an ideal for many years, yet there is little evidence that the fully automated, flexible manufacturing facilities discussed in concept as early as the 1960s are a reality today [10]. The practical problems which have delayed the implementation of this concept relate not only to technological difficulties, but also to investment limitations and to the need to keep an operation functioning during the changeover. This has resulted in 'islands of automation' addressing aspects of the operation where costs of traditional processes are high and returns are relatively short-term. These 'islands' may be linked via software to allow data to be transferred automatically between one manufacturing stage and another.

Successful CIM is more than the automation of traditional processes. It offers best returns if the whole operation of the business — including product design, provisioning of materials, manufacturing operations, planning and control are integrated via a computerized information system. Implementation of office automation systems has demonstrated that it is not sufficient to convert 'paper' messages to 'electronic mail'. Departments get overloaded with information in whatever form it is transmitted unless the communication systems and processes are carefully planned and controlled so that only relevant and necessary information is transmitted in a compact and efficient form.

As part of the British government-backed Alvey programme of artificial intelligence based projects, GEC Electrical Projects are leading a research team investigating 'design-to-product' capability. The project, which commenced in 1984, will span five years at a cost of £8.5 million.

One concept of CIM involves a hierarchical structure of automation. For example, at the first level of integration a range of automatic test instruments may operate under the control of a computer to provide a test station (Fig. 9.1). The next level involves the overall coordination of a complete test department, involving ICT and FNT systems as well as fault diagnosis and repair stations. At level 3 the process flows throughout the manufacturing department are controlled by computer, while at the highest level the corporate management system receives inputs and provides overall strategic planning and control.

There are dangers, however, in accepting the traditional structure of an electronics business as the basis for CIM. In previous chapters the concepts of 'feedforward' of design data to assembly, test and repair have been discussed. The process, whilst avoiding the recreation of initial data, is still essentially open-loop in nature. For efficient control an automatic closed-loop system is required, and this involves 'feedback'. Faults detected at each stage in the manufacturing process, for example, should influence the design database. When these faults can be quickly corrected in manufacture, the only effect may be an update to statistical records which can be used as the basis for selection of components or processes in future designs. In this way quality control comes to mean much more than detection and prevention of delivery of faulty product to the customer, but rather a means of building up a more reliable design / manufacturing system — the environment within which future products will be defined, designed and realized.

Networks designed for this type of role are emerging. An example of such a facility is

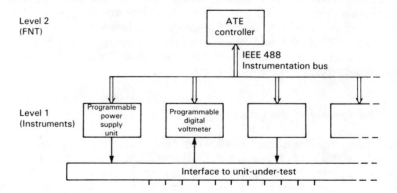

Fig. 9.1

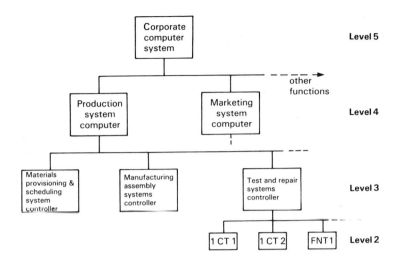

Fig. 9.2

GRnet from GENRAD. GRnet is a high-speed data network specifically designed for automatic test and repair systems, and is used in GENRAD's TRACS (Test and Repair Analysis/Control System). Data from each workstation or automatic test /repair equipment are collected; diagnostic messages are transferred between stations; and analysis and management reports are prepared automatically.

The upgrading of such systems to levels four and five (Fig. 9.2) may occur rapidly now that large companies, such as IBM who have pioneered CIM, are setting up divisions specifically to address this requirement. International standards in this area are urgently required if the efforts of individual pioneering companies are not to be largely dissipated.

General Motors have established a set of baseline standards, which they call the Manufacturing Automation Protocol (MAP), designed to guide companies supplying products to be used as 'islands of automation' within General Motors' factories. A number of electronics component and computer manufacturers — including Intel and Tandem — have announced their support of the MAP standard.

9.4.3 Expert systems and the artificial intelligence barrier

Already system suppliers are labelling some of their latest CAD/CAM products as 'expert systems'. For example, MSDI has recently launched modules for sheet metal fabrication and surface modelling which it claims acquire applications expertise from users of the system.

Future generations of CAE systems will incorporate artificial intelligence features whereby the rules are developed by the system itself based upon reasoning and experience [11,12]. This degree of flexibility should ultimately lead to the availability of 'product synthesizers' which not only are capable of designing and producing an optimum solution to a defined operational requirement, but which may also be used to postulate new concepts — and hence to 'create markets' and the criteria for their realization in optimized form.

SUMMARY

Further advances in CAE systems will benefit from continued improvements in component technology, particularly in the areas of VLSI silicon circuits, which will probably remain the major realization medium for electronics products for several years. Gallium Arsenide technology is maturing rapidly, with currently 10^5 devices on a chip representing the state of the art in integration. The superior speed performance of GaAs will ensure that it finds increasing application not only in critical analogue signal processing applications, but also in real-time digital systems such as machine vision, speech input/output and precision manipulation which involve rapid processing of large amounts of data.

Limitations of the Von Neumann computer architecture are recognized, and super-computers employing parallel processing may soon become apparent in real-time CIM systems. At present two routes to total system integration are emerging. The moves by large companies such as IBM and Hewlett Packard towards extension of their product ranges to embrace sub-systems which address each element of the product life cycle from design through materials provisioning, process control, manufacture, test and customer support may ultimately lead to general availability of their CIM systems to the industry. Alternatively, other CAE system vendors and users will have to work more closely together to define and maintain international standards — such as EDIF — which will permit the smaller specialist CAE companies to concentrate upon their areas of expertise whilst allowing OEMs to achieve efficient integration of a range of tools and communication systems selected to meet their own particular needs.

REFERENCES

1. Bennets, R.G. (1984) *Design of Testable Logic Circuits*. Addison-Wesley, Wokingham.
2. Easterbrook, J. (April 1985) BIST — Built-in self test. *Silicon Design*, **2**(4).
3. Frate, J. (Ed.) (June 1985) Gallium arsenide ICs come out of the shadows. *New Electronics*, **18**, 27–37.
4. Firstenberg, A. and Roosild, S. (1985) GaAs ICs for new defense systems offer speed and radiation hardness benefits. *Microwave Journal*, 145–63 (March).
5. Richardson, M. (1985) Parallel applications. *Systems International*, **13**(8), 41–2.
6. Weiss, R. (1985) Computer architectures: varied blueprints will lift speeds to dizzying heights. *Electronic Design*, 83–92 (May).
7. Meijer, A. and Peeters, P. (1982) *Computer Network Architectures*. Pitman, London.
8. Feur, M. (July 1985) Semicustom vendor interfaces: a CAE perspective. *New Electronics*, **18**(15), 102.
9. Stover, R.M. (1984) *An Analysis of CAD/CAM Applications*. Prentice-Hall, Englewood Cliffs, New Jersey.
10. Dance, M. (March 1985) Industrial automation — the factory of the future. *Electronics Industry*, **11**(3), 9–19.
11. Begg, V. (1984) *Developing Expert CAD Systems*. Kogan-Page, London.
12. Gevarter, W.B. (1985) *Intelligent machines*. Prentice-Hall, Englewood Cliffs, New Jersey.

Appendix 1
Learning objectives

The general and specific learning objectives listed below are offered as the basis for the creation of a syllabus centred upon the contents of this book. All objectives are intended to be prefaced by the expression 'The expected learning outcome is that the student . . .'

A EVOLUTION OF COMPUTER AIDED ELECTRONIC ENGINEERING

1 Appreciates the major developments which have contributed to CAE technology.

1.1 Discusses the work of pioneers in computer technology and graphics, and the significance of their contributions to the development of CAE systems.

1.2 Compares and contrasts the characteristics of modern computer graphics input and output devices.

1.3 Calculates screen memory requirements for various monochrome and colour display systems.

2 Understands the background to CAE and its significance throughout the life cycle of an electronic product.

2.1 Discusses the pressures which have led to a compression of product life cycles.

2.2 Outlines aspects of the development, manufacture and support of a product which can be addressed by CAE.

2.3 Lists and briefly explains the main steps in:
(a) a development project
(b) manufacture of an SIC
(c) manufacture of a hybrid microcircuit
(d) manufacture of a PCB assembly.

2.4 Briefly discusses customer-orientated aspects of running an electronic engineering business, and outlines aspects which CAE can be used to address more effectively.

B COMPUTER AIDED DESIGN

3 Understands the principles of circuit simulation.

3.1 Discusses the concepts of analysis, simulation, synthesis and optimization and relates them to modern computer aided design systems.

3.2   Discusses the main facilities and models within a modern circuit simulator (e.g. SPICE).
3.3   Sets up and performs a circuit simulation to establish the d.c., a.c., harmonic, transient, and noise performance (as appropriate) of:
        (a) a small signal linear circuit
        (b) a non-linear circuit
        (c) a logic circuit.
        using a circuit-level simulator.
3.4   Explains the justification for selective rather than exhaustive testing of large-scale logic systems.
3.5   Calculates the number of test vectors to exhaustively test a logic system containing combinational logic and latches.
3.6   Derives a reduced test vector set for a simple logic circuit using the 'fault matrix' technique.
3.7   Develops a description of a logic system using a hardware description language (HDL).
3.8   Sets up and runs a logic simulation of a circuit from its HDL description.
3.9   Uses a logic simulator to determine the degree of fault coverage provided by a set of test vectors.

4   Appreciates the operational features of a modern CAD system.
4.1   Creates an input file for a CAD system either as a net list or via a schematics capture package.
4.2   Devises a satisfactory placement and routing scheme using the interactive editor for all or part of the design.

5   Understands the commercial and technical features of standard cell and gate array integrated circuits.
5.1   Compares standard cell and gate array SICs in terms of manufacturing procedures, chip-area utilization, development timescale and cost, and manufacturing cost as a function of circuit complexity and production quantity.
5.2   Discusses the problems of cell placement for standard cell and gate array SICs.
5.3   Outlines the principle of various algorithms for autorouting of semicustom integrated circuits and compares their performance in terms of routing success, memory requirements and speed.
5.4   Explains the aims of a silicon compiler with reference to typical systems in current use.

6   Appreciates the characteristics of modern CAD systems.
6.1   Compares typical micro- mini- and workstation based CAD systems in terms of performance and cost.
6.2   Creates a set of manufacturing data for a PCB assembly using a modern CAD system.
6.3   Discusses the significance of standards for operating systems and networking (e.g. UNIX and Ethernet) to CAD systems developments.
6.4   Compares the key performance features of various types of printers and plotters for CAD applications, and hence selects appropriate equipments for particular applications.

C   COMPUTER AIDED MANUFACTURING

7   Understands the processes involved in manufacture and assembly of PCBs.
7.1   Sketches a flow chart for PCB manufacture and explains the key steps.
7.2   Differentiates between NC, CNC, and DNC machining operations.
7.3   Prepares a double-sided PCB using data produced from the CAD system for primary and secondary masking.
7.4   Compares manual, semi-automatic and automatic assembly of PCBs (using inserted and surface mounted components) in terms of accuracy, throughput and capital cost.
7.5   Discusses various soldering processes and the technical and commercial criteria for selection of a process.

D COMPUTER AIDED TEST

8   Understands the factors which determine component test philosophy.
8.1   Explains the relationship between optimum test strategy and fault spectrum.
8.2   Calculates first pass success probabilities in terms of component quality and assembly complexity.
8.3   Evaluates the relative costs of zero, sample, and 100% goods inwards testing of components from data on fault probabilities and repair costs at various stages in the product life cycle.
8.4   Calculates probability of field failures as a function of component 'infant mortality life-times', and burn-in temperature and duration.

9   Understands the principles of in-circuit and functional testing.
9.1   Discusses the principle and practical problems of in-circuit and functional testing.
9.2   Selects a suitable test mix and test strategy given data on first-pass yield and fault coverage for each type of test station.
9.3   Sets up a simple functional ATE using a microcomputer with IEEE488 interface and controllable instrumentation.
9.4   Devises a test schedule for an electronic assembly and converts it to a program to run on the ATE.

10   Appreciates the factors which influence testability of a VLSI devices both singly and on PCB assemblies.
10.1   Explains the influence of controllability and observability of circuit nodes upon the testability of a device.
10.2   Explains the principle of 'scan path' and 'built-in self test' techniques for improving the testability of circuits involving sequential logic.
10.3   Discusses various design features which contribute to the testability of a VLSI PCB assembly.

11   Appreciates performance/cost trade-offs in typical ATE systems.
11.1   Creates a comparison table summarizing the cost and key performance features of a number of component and PCB testers.
11.2   Discusses the technical, financial and communications implications of integrating design and test via a common database.

E MANAGEMENT SYSTEMS

12   Understands product and project database concepts.
12.1   Explains how several product databases can reside within a company database, and discusses the authorization controls which must be established.

13   Understands project planning and control techniques.
13.1   Uses PERT and CPA techniques to plan assignment or project work, establishing milestones and estimating resource requirements.
13.2   Inputs activities and dependency data to a project planning software package, and obtains network and/or bar-chart plans.
13.3   Derives, via the project plan, earliest and latest start and finish dates for each work package.
13.4   Derives a line of balance for a repetitive process within a project plan, and hence predicts schedule variance at completion.
13.5   Discusses the benefits of linking technical and temporal/cost management systems via a common database.
13.6   Prepares a project review report (on an assignment or project during implementation) using the 'earned value' concept.

14   Appreciates the key factors involved in appraisal of investments in CAE.

14.1   Outlines the principles of:

(a) pay back

(b) return on investment

(c) discounted cash flow

criteria in investment appraisal.

14.2   Creates a decision base for selection of CAE equipments and sets up a spreadsheet for its analysis.

14.3   Compares two CAE projects covering different timescales using DCF return on investment.

15   Appreciates human factors important to the successful implementation of a CAE project.

15.1   Discusses factors which are likely to influence the 'acceptability' of CAE investments by the people who are to operate the systems.

15.2   Discusses the benefits of training in advance of system installation and considers options for its implementation.

16   Recognizes the need for continued development of computer aided electronic engineering systems.

16.1   Discusses technological trends in components and in hardware and software systems which are likely to influence the requirements and the development of CAE systems.

16.2   Explains the concept of computer integrated manufacturing (CIM) and 'product synthesizers', and the capabilities and limitations of current systems aimed towards these goals.

Index